South University Library
Richmond Campus
2151 Old Brick Road
Glen Allen, Va 23060

FEB 0 2 2011

HOLISM AND EVOLUTION

BY
GENERAL THE RIGHT HONORABLE
J. C. SMUTS

New York
THE MACMILLAN COMPANY
1926

All rights reserved

COPYRIGHT, 1926,
BY J. C. SMUTS.

Set up and printed.
Published September, 1926.

Printed in the United States of America by
J. J. LITTLE AND IVES COMPANY, NEW YORK

PREFACE

THIS work deals with some of the problems which fall within the debatable borderland between Science and Philosophy. It is a book neither of Science nor of Philosophy, but of some points of contact between the two. To my mind it is the surface of contact between the two that will prove fruitful and creative for future progress in both, and to which special attention should be directed. Some border problems between the two are here considered in the light of recent advances in physical and biological science. And a re-examination of fundamental concepts in the light of these advances reveals the existence of a hitherto neglected factor or principle of a very important character. This factor, called Holism in the sequel, underlies the synthetic tendency in the universe, and is the principle which makes for the origin and progress of wholes in the universe. An attempt is made to show that this whole-making or holistic tendency is fundamental in nature, that it has a well-marked ascertainable character, and that Evolution is nothing but the gradual development and stratification of progressive series of wholes, stretching from the inorganic beginnings to the highest levels of spiritual creation. This work deals with our primary concepts of matter, life, mind and personality in the light of this principle, and discusses some of the problems of Evolution from this new point of view. The discussion is not technical, specialist or exhaustive in any sense. It is intended to sketch and explain the general lines of argument rather than to go into details. It is especially the fundamental concept of Holism which I wish to explain and justify, as well as the scientific and philosophic viewpoint to which it leads. The detailed elaboration must be left to more competent hands and to those favoured with more leisure than I can find in a busy public life. I have tried to sketch the general lines of reasoning in a way which, while I hope scientifically accurate so far as they go, are yet popular enough to be readily understood by readers with a fair average reading in general science.

It is my belief that Holism and the holistic point of view will prove important in their bearings on some of the main problems of science and philosophy, ethics, art and allied subjects. These bearings are, however, not fully discussed in this work, which is more of the nature of an introduction, and is concerned more with the laying of foundations than with the superstructure. I have no time at present to do more than write an introductory sketch; but I hope in the years to come to find time to follow up the subject and to show how it affects the higher spiritual interests of mankind. The old concepts and formulas are no longer adequate to express our modern outlook. The old bottles will no longer hold the new wine. The spiritual temple of the future, while it will be built largely of the old well-proved materials, will require new and ampler foundations in the light of the immense extension of our intellectual horizons. This little book indicates the lines along which my own mind has travelled in the search for new and more satisfactory concepts.

A generation ago, when I was an undergraduate at Cambridge, the subject of Personality interested me greatly, and I wrote a short study on "Walt Whitman: A Study in the Evolution of Personality," in order to embody the results I had arrived at. This study was never published, but the subject continued at odd intervals to engage my attention. Gradually I came to realise that Personality was only a special case of a much more universal phenomenon, namely, the existence of wholes and the tendency towards wholes and wholeness in nature. In 1910 I sought relief from heavy political labours in an attempt to embody my new results in a study called " An Inquiry into the Whole," which also was not published. I had no time to return to the subject until, in 1924, a change of government released me from burdens which I had continuously borne for more than eighteen years. When I came to read once more the

MS. of fourteen years earlier I found much of the scientific setting out of date and I found my conception of Holism had also altered in certain respects. I therefore decided once more to make a fresh start with my study of wholes and Holism in nature. The present work is the first-fruits of this fresh effort. The aspects and bearings of Holism in which I am mainly interested are not yet reached in this study, which, as I have said, is of an introductory character. But I feel that unless I now make a determined attempt to prepare at least a part of my inquiry for publication, it will in all probability never get beyond the incubation stage in which it has remained so many years. This I would personally regret, as I think that in Holism we have an idea which may perhaps prove valuable and fruitful, and which for. better or worse should be lifted out of the obscurity in which it has so long remained in my mind. Whether my partiality for the idea, which has been my companion throughout a crowded life, will be shared by others, time alone will show.

The work has unfortunately had to be written in somewhat of a hurry and amid the pressure of many other calls on my time. Nor in writing it have I had the advantage of consulting any expert friends on details. This must be my excuse for any incidental mistakes or slips which may be found in it.

J. C. SMUTS.

Irene, Transvaal,
September 1925.

CONTENTS

CHAPTER		PAGE
	Preface	v
I.	The Reform of Fundamental Concepts	1
II.	The Reformed Concepts of Space and Time	22
III.	The Reformed Concept of Matter	35
IV.	The Cell and the Organism	59
V.	General Concept of Holism	85
VI.	Some Functions and Categories of Holism	118
VII.	Mechanism and Holism	145
VIII.	Darwinism and Holism	182
IX.	Mind as an Organ of Wholes	224
X.	Personality as a Whole	261
XI.	Some Functions and Ideals of Personality	290
XII.	The Holistic Universe	317
	Index	347

HOLISM AND EVOLUTION

HOLISM AND EVOLUTION

CHAPTER I

THE REFORM OF FUNDAMENTAL CONCEPTS

Summary.—In spite of the great advances which have been made in knowledge, some fundamental gaps still remain; matter, life and mind still remain utterly disparate phenomena. Yet the concepts of all three arise in experience, and in the human all three meet and apparently intermingle, so that the last word about them has not yet been said. Reformed concepts of all three are wanted. This will come from fuller scientific knowledge, and especially from a re-survey of the material from new points of view. The fresh outlook must accompany the collection of further detailed knowledge, and nowhere is the new outlook more urgently required than in the survey of these great divisions of knowledge.

Take Evolution as a case in point. The acceptance of Evolution as a fact, the origin of life-structures from the inorganic, must mean a complete revolution in our idea of matter. If matter holds the promise and potency of life and mind it is no longer the old matter of the physical materialists. We have accepted Evolution, but have failed to make the fundamental readjustment in our views which that acceptance involves. The old mechanical view-points persist, and Natural Selection itself has come to be looked upon as a mere mechanical factor. But this is wrong: Sexual Selection is admittedly a psychical factor, and even Natural Selection has merely the appearance of a mechanical process, because it is viewed as a statistical average, from which the real character of struggle among the living has been eliminated.

Nineteenth-century science went wrong mostly because of the hard and narrow concept of causation which dominated it. It was a fixed dogma that there could be no more in the effect than there was in the cause; hence creativeness and real progress became impossible. The narrow concept of causation again arose from a wider intellectual error of narrowing down all concepts into hard definite contours and wiping out their indefinite surrounding "fields." The concept of "fields" is absolutely necessary in order to get back to the fluid plastic facts of nature. The elimination of their "fields" in which things and concepts alike meet and intermingle creatively made all understanding of real connections and inter-actions impossible. The double mistake of abstraction and generalisation has thus led to a departure in thought from the fluid procedure of nature. This narrowing of concepts and processes into hard and rigid outlines, and their rounding off into definite scientific contours temporarily simplified the problems of science and thought, but we have outlived the utility of this procedure, and for further advance we have now to return to the more difficult but more correct view of the natural plasticity and fluidity of natural things and processes. From this new view-point a re-survey will be made in the sequel of our ideas relating to matter, life and mind, and an attempt will be made to reach the fundamental unity and continuity which underlie and connect all three. We shall thus come to see all three as connected steps in the same great Process, the nature and functions of which will be investigated.

AMONG the great gaps in knowledge those which separate the phenomena of matter, life and mind still remain unbridged. Matter, life and mind remain utterly unlike each other. Apparently indeed their differences are ultimate, and nowhere does there appear a bridge for thought from one to the other. And their utter difference and disparateness produce the great breaks in knowledge, and separate knowledge into three different kingdoms or rather worlds. And yet they are all three in experience, and cannot therefore be so utterly unlike and alien to each other. What is more, they actually intermingle and co-exist in the human, which is compounded of matter, life and mind. If indeed there were no common basis to matter, life and mind, their union in the human individual would be the greatest mystery of all. What is in fact united in human experience and existence cannot be so infinitely far asunder in human thought, unless thought and fact are absolutely incongruous. Not only do they actually co-exist and mingle in the human, they appear to be genetically related and to give rise to each other in a definite series in the stages of Evolution; life appearing to arise in or from matter, and mind in or from life. The actual transitions have not been observed, but are assumed to have taken place under certain conditions in the course of cosmic Evolution. Hence arise the three series in the real world: physical, biological and psychical or mental. These connections between them, which are based not on thought but the facts of existence and experience, tend to show that they cannot be fundamentally alien and irreconcilable, and that some sort of bridge between them must be possible, unless we are to assume that our human experience is indeed a mere chaotic jumble of disconnected elements.

As I have said, the problem does not arise from the facts either of experience or of existence. The problem is one for our thought and our science. It is for our thought that the mystery exists, and it is for knowledge that the great gaps between the physical, the biological and the mental series arise. The solution must therefore ultimately depend on our more extended knowledge of these series and the discovery of interconnections between them. The great darknesses and gaps in experience are mostly due to ignorance. Our experience is clear and luminous only at certain points which are separated by wide regions of obscurity; hence the apparent mystery of the luminous points and of their isolation and unlikeness. Hence also the still greater mystery of their actual union in the threefold incarnation which constitutes human personality.

But it is just this union which ought to warn us that the apparent separateness of these three fundamental concepts is not well founded in fact, and that a wider knowledge and a deeper insight might be able to clear up the mystery, at least to some extent, and to produce for thought as for existence a sort of union or harmony of these apparently unrelated or independent elements in our real world. More knowledge is wanted. Our physical science ought to provide the solvent for our idea of hard impenetrable inert matter, and in the third chapter I shall inquire in how far there are already the materials for such a solvent. Again, our biological science should dispel the vagueness of the concept of life, and replace it by a more definite meaningful concept, which will yet not depend on purely material or physical elements. At present the concept of life is so indefinite and vague that, although the Kingdom of life is fully recognised, its government is placed under the rule of physical force or Mechanism. Life is practically banished from its own domain, and its throne is occupied by a usurper. Biology thus becomes a subject province of physical science—the Kingdom of Beauty, the free artistic plastic Kingdom of the universe, is inappropriately placed under the iron rule of force. Mind again, which is closest to us in experience, becomes farthest from us in exact thought. The concepts in which we envisage it are so vague and nebulous, compared with the hard and rigid contours of our concepts of matter, that the two appear poles asunder. Here too a reformed concept of mind might bring it much closer to a reformed concept of matter. And thus, out of the three at present utterly heterogeneous polar concepts of matter, life and mind it might be possible to develop concepts moulded more closely to fact and experience, freed of all adventitious and unnecessary elements of separateness and disparity, and forming (as in all true science they should rightly form) the co-operative elements and aspects in a wider, truer conception of Reality. It may be said that in making this demand for new concepts of matter, life and mind we are imposing an impossible task on thought. We are asking it to go beyond itself and deal with matters entirely beyond its own proper world. Matter, it may be urged, is essentially outside and beyond thought, something hard and impervious to thought, an object to thought which thought can only just barely reach up to in its utmost effort, and no more. Life is, of course, not alien to thought in the same sense as matter, but still it also falls outside the province of thought, it also has a reality of its own beyond thought, and it also is an object to thought. How then could thought embrace these provinces, how could it be a measure of these provinces beyond its ken; how could the part envisage the whole? Our standard of measurement is inadequate, our task therefore impossible.

The answer is that, while mind or thought may not have made matter, it has undoubtedly assisted in making the concept of matter; and this concept, based as it is on mere empirical experience, on inadequate knowledge and covered with a thick over-burden of unsifted tradition, may be mostly wrong, however deeply embedded in human thought it may be. A reform of the concept of matter is urgently required, and is indeed amply justified by the unprecedented recent advances in physical science, and especially in our knowledge of the constitution of matter. And a reform will, as I shall show in the third chapter, bring matter considerably nearer to the concept of life.

With regard again to the concept of life, what is most urgently required is that it should be rid of that haziness, indefiniteness, and vagueness which makes it practically worthless for all exact scientific purposes. Biological science has not in recent years made the same gigantic strides forward in the knowledge of fundamentals that physical science has taken, and yet for Biology too the sky has considerably cleared, and what two or three decades ago was still hotly disputed is to-day generally accepted. Besides, the greatest development in Biology during this century has taken place in the science of Genetics, and the trend there has been steadily away from the hard mechanical conceptions which dominated Biology more than a generation ago. The time here too may be ripe for a reconsideration of some of the fundamental concepts and standpoints. I may express the hope that the masters of this science will not concentrate all their attention on special researches, however promising the clues at present followed may be; but that they will find time for a reconsideration of the wider conceptions which, ever since the great time of Darwin, have been getting further out of gear. Unless Biology can succeed in clarifying her fundamental conceptions there is risk of great confusion in a science in which old general ideas have persisted in spite of great progress in detailed knowledge. If in the sequel I join in the discussion of the foundations of Biology, not as entitled of right to speak but more in the character of an outside spectator urging the importance of a new point of view, I hope my presumption in so doing may be forgiven me.

For welcome as any new and deeper knowledge would be on these high matters, the present situation calls even more urgently for fresh points of view. Matter, life and mind are, so to speak, the original alphabet of knowledge, the original nuclei of all experience, thought, and speculation. Their origin is purely empirical, their course has been shaped by tradition for thousands of years, and all sorts of discarded philosophies have gone towards the making of their popular meanings. In spite, therefore, of the great fundamental aspects of truth which they embody, the kernel of truth in them has become overlaid by deep incrustations of imperfect and erroneous knowledge. Modern science and philosophy have repeatedly ventured on reforms, but the popular use of these terms tends to obliterate all fine distinctions. I do not believe that an abiding scientific or philosophic advance in this respect will be possible until a more exact nomenclature has been adopted. Such a reform I am going to advocate and suggest in the sequel, but in the meantime I wish to emphasise how important it is, not merely to continue the acquisition of knowledge, but also to develop new view-points from which to envisage all our vast accumulated material of knowledge. The Copernican revolution was not a revolution in the acquisition of new knowledge, but in view-point and perspective in respect of existing knowledge. The most far-reaching revolutions in knowledge are often of this character. Evolution in the mind of Darwin was, like the Copernican revolution, a new view-point, from which vast masses of biological knowledge already existing fell into new alignments and became the illustrations of a great new Principle. And similarly Einstein's conception of General Relativity in the physical universe, whatever its final form may yet be, is a new view-point from which the whole universe and all its working mechanisms acquire a new perspective and meaning.

More knowledge is undoubtedly required, but its acquisition must go hand in hand with the exploration of new concepts and new points of view. It will not help merely to accumulate details of which, even in the special departments of the separate biological sciences, the masses are already becoming more than any individual mind can bear. New co-ordinations are required, new syntheses which will sum up and explain and illuminate the otherwise amorphous masses of material. While research is being prosecuted as never before, while in biological science great, and in the physical sciences unprecedented, progress is being recorded, the call becomes ever more urgent for a reconsideration of fundamental concepts and the discovery of new standpoints which might lead to the formulation of more general principles and wider generalisations. Nowhere are new viewpoints more urgently called for than in respect of the fundamental concepts of matter, life, and mind, of which the reform is overdue and the present state is rapidly becoming a real obstacle to further progress. And I may point out that the formulation of new view-points will depend not so much on masses of minute details, as on the consideration of the general principles in the light of recent advances, the collation and comparison of large masses of fact, and the survey of fairly large areas of knowledge. The road is to be discovered, not so much by minute local inspection as by wide roaming and exploration and surveying over large districts. Both methods are needed, and the question narrows itself down to one of comparative values. Just as in the cases of Newton and Einstein, the new clues are more likely to be indicated by certain crucial dominant facts than by small increments of research. It would therefore be a great mistake to let the completion of present detailed researches take precedence over the more general and urgent questions to which I am drawing attention.

Let me mention one matter of crucial significance to which I think sufficient importance has not yet been attached. To-day I think it is generally accepted that life has in the process of cosmic Evolution developed from or in the bosom of matter, and that mind itself has its inalienable physical basis. I do not think that among those who have given thought and attention to these matters there are to-day any who seriously question this position. Life is no dove that has flown to our shores from some world beyond this world; mind or soul is not an importation from some other universe. Life and mind are not mere visitants *to* this world, but not *of* this world. There is nothing alien in them to the substance of the universe; they are with us and they are of us. The popular view still looks upon the association of life and mind with matter as a sort of symbiosis, as the close living together of three different beings, as the dwelling of life and the soul in the body of matter, just as in the organic world one plant or animal organism will be found normally living with and in another. This popular traditional view comes from the hoary beginnings of human thought and speculation, but it is definitely abandoned by all those who have assimilated the modern view-point of Evolution. For them in some way not yet fully understood, but accepted as an undoubted fact, both life and mind have developed from matter or the physical basis of existence. The acceptance of this fact must have far-reaching consequences for our world-view.

But before I refer to that aspect of the matter let me point out how this acceptance affects the grave issues over which our fathers fought a continuous battle royal during the latter half of the nineteenth century. The materialists contended for this very point, namely, that life and mind were born of matter. From this they proceeded (quite illegitimately) to infer the primacy and self-sufficiency of matter in the order of the universe, and to reduce life and mind to a subsidiary and subordinate position as mere epiphenomena, as appearances on the surface of the one reality, matter. To use the Platonic figure, to them matter was the lyre, and the soul was the music of that lyre; the lyre was the substantive and abiding reality, and the music a mere passing product. And thus the priority and dominance of matter made of life and the soul merely transient and embarrassed phantoms on the stage of existence. This materialism was most hotly resented and contested by those who held to the spiritual values and realities. They denied not only the primacy of matter but also that life or mind sprang from it and were dependent on it in any real sense. In fact they denied the principle of Evolution as undermining all the spiritual and moral values of life. Both sides, materialists and spiritualists alike, were under the influence of the hard physical concepts of cause and effect which played such a great part in the science of the nineteenth century. There could be nothing more in the effect than there was already in the cause; and if matter caused the soul, there could be nothing more in the soul than there already was in matter. In other words, the soul was merely an apparent and no real substantial advance on matter. The abstract validity of this argument was never questioned and was thoroughly believed in by both sides. Hence those who affirmed the theory of Evolution logically tended to be materialists, and those who were spiritualists were logically forced to deny Evolution.

Without their knowing it the great battle raged, not over the facts of Evolution, but over a metaphysical theory of causation in which they both believed and were both wrong. Such is the irony of history. To-day we pick the poppies on the old bloody battlefield of Evolution, and can afford to be fair to both sides. The essential terms have changed their meaning for us. We believe in Evolution, but it is no more the mechanical Evolution of a generation or two ago, but a creative Evolution. We believe in the growth which is really such and becomes ever more and more in the process. We believe in Genesis which by its very nature is epigenesis. For us there is no such thing as static Evolution, a becoming which does not become but in its apparent permutations ever remains the same. The absolute equation of cause and effect, which was a dogma implicitly believed in by the men of that day, does not hold for us, as I shall later on explain. The temperature has changed, the view-point has shifted, and to-day thoughtful men and women are sincere and convinced Evolutionists, without troubling themselves over the dead and forgotten issue of materialism versus spiritualism. We accept the theory of descent, of life from matter, and of the mind from both. For educated men and women to-day Evolution is just as much part and parcel of their general outlook, of their intellectual atmosphere, so to say, as is the Copernican theory.

As I said before, this is a fact with very far-reaching implications. If we believe that life and mind come from matter, if they are evolved from matter, if matter holds the promise, the dread potencies of life and mind, it can for us no longer be the old matter of the materialists or the physicists. The acceptance of the view for which the materialists fought so hard means in effect a complete transformation of the simple situation which they envisaged. Matter discloses a great secret; in the act of giving birth to life or mind it shows itself in an entirely unsuspected character, and it can never be the old matter again. The matter which holds the secret of life and mind is no longer the old matter which was merely the vehicle of motion and energy. The landmarks of the old order are shifting, the straight contours of the old ideas are curving, the whole situation which we are contemplating in the relations of matter, life and mind is becoming fluid instead of remaining rigid. The point to grasp and hold on to firmly is that the full and complete acceptance of Evolution must produce a great change in the significance of the fundamental concepts for us. Life and mind now, instead of being extraneous elements in the physical universe, become identified with the physical order, and they are all recognised as very much of a piece. This being so, it obviously becomes impossible thereupon to proceed to erect an all-embracing physical order in which life and mind are once more declared aliens. This cat and mouse procedure is simply a case of logical confusion. This in-and-out game will not do. If Evolution is accepted, and life and mind are developments in and from the physical order, they are in that order, and it becomes impossible to continue to envisage the physical order as purely mechanical, as one in which they have no part or lot, in which they are no real factors, and from which they should be logically excluded. If

Evolution is right, if life and mind have arisen in and from matter, then the universe ceases to be a purely physical mechanism, and the system which results must provide a real place for the factors of life and mind. To my mind there is no escape from that argument, and its implications must have a very far-reaching effect on our ideas of the physical order, and on a biology in which mechanical views are still dominant.

The point I have been trying to make is that our ultimate concepts need reconsideration, and that above all new view-points are necessary from which to re-survey the vast masses of physical and biological knowledge which have already accumulated. I have said that certain large dominant facts may be sufficient to lead to a new orientation of our ideas. And I have taken the accepted fact of Evolution as a case in point. The older materialists and the present-day mechanical biologists have both fought hard for the acceptance of Evolution as a fact, without realising that such an acceptance must inevitably mean a transformation of their view-points, and that both the meaning of the concept of matter and the idea of the part played by mechanism in biology must be seriously affected by such acceptance. It is clear that the full significance of the great dominant idea of Evolution and its effect on the ordering of our ultimate world-view are not yet fully realised, and that we are in effect endeavouring simultaneously to go forward with two inconsistent sets of ideas, that is to say, with the idea of Evolution (not yet adequately realised) and the pre-Evolution physical ideas (not yet quite abandoned). This is, however, sheer confusion, and a clarification of our ideas and the realisation of new view-points have become necessary.

Let me now leave the general fact of Evolution as bearing on our world-view and call attention to another and somewhat similar case which arises in Darwin's theory of Descent. In that theory Natural Selection is usually but erroneously taken to be a purely mechanical factor. It is understood to operate as an external cause, eliminating the unfit in the struggle for existence, and leaving the fit in possession of the field to reproduce their kind and to continue the story of Evolution. Natural Selection, from whatever cause arising and in whatever way operating, is on this view taken to be merely a mechanical cause or factor, just as is a hailstorm which kills plants and animals with hailstones, or a drought which kills them from want of water. Whether the destruction arises from physical mechanical causes like storms or drought or lightning, or whether it arises from the action of living agencies such as enemies or disease or the like, makes no difference to the result, which is in either case the same. A broad generalised statistical view of the causes of elimination has been taken and Natural Selection as a whole has on this view assumed the appearance of a mere external mechanical factor. On this view of Natural Selection, therefore, Darwin's theory of Evolution has come powerfully to reinforce the generally prevalent mechanical position. The effect has been just the opposite to what one might have expected from a great biological advance. The kingdom of life, instead of fighting for its own rights and prerogatives, has tamely and blindly surrendered to the claims of physical force and actually joined hands with it and contributed to its supremacy. The acceptance of Darwinism, therefore, so far from stemming the tide of mechanical ideas, has actually furthered and assisted it, and raised it to full flood. Through a misconception of the nature of Natural Selection the mechanical ideas have invaded the domain of life, where opposition might have been expected, and through the conquest and occupation of that domain the mechanical position, which would otherwise have been confined to the material physical sphere, has been extended and powerfully consolidated. This result was due, as I say, to the generalised statistical concept of Natural Selection, irrespective of whether it was due to organic or inorganic causes.

There is, however, one form of Selection which cannot be thus indiscriminately dealt with. It arises not only from organic causes, but still more narrowly and quite indisputably from psychical or mental causes. Darwin called it Sexual Selection, and in spite of the opposition of A. R. Wallace and others not only attributed great importance to it, but as time went on and he saw his great vision more clearly, he gave it an ever-growing emphasis in his theory of Descent. Natural Selection operates on the unfit by destroying them or killing them off; Sexual Selection, on the contrary, has a more limited operation and applies only in respect of males, whose reproduction it handicaps, limits or prevents. In other words it is a struggle among males for the possession of females, and in this struggle males are assisted not only by their superior strength or fighting powers, but also by their superior power of song or beauty or scent or general attractiveness or excitiveness to females. It is clear that the real motive power of this form of selection is mostly biological and psychical. The female is excited and attracted by superior fighting force or superior artistic endowments among males competing for her favour. And when one considers the degree of perfection to which the male forms have attained largely under this stimulus of the female sex instinct, one is struck with amazement at the emotional sensitiveness thus implied on the part of female insects, birds, and beasts, and at the wonderful subtlety and fineness of the emotional discrimination which has resulted from it. The beauty of form and colour which characterises, for instance, the peacock's feathers are such that even our human eye can scarcely do justice to it. And yet on the principle of Sexual Selection that perfection of beauty is due to the amazing emotional sensitiveness and appreciation of the peahen, which through countless generations must have been attracted by the minute superiority of the one male over others in this respect. And the same applies in regard to the wonderful power of song among male birds and all the other secondary male characters. The psychical emotional powers implied on the part of the female on this theory are so wonderful as to be almost unintelligible; in many respects they are superhuman, and would appear to throw an astonishing light on the unconscious psychical developments of what we are pleased to call the lower creatures. If secondary sexual characters with all their perfection of form and colour did originate and develop under the stimulus of Sexual Selection, we shall have to revise our views radically as to the psychical sensitiveness and endowments of large classes of these lower animals. If, for instance, the human female showed the same sensitive judgment and discrimination as the female bird or insect shows a sensitive sex instinct, what supermen we sorry males in time would become! I am afraid, however, that the theory of Sexual Selection does not tell the whole story, and that there is more in the sexual situation than appears from that theory. I shall refer to the subject again in Chapter VIII. But for my present purpose it is only necessary to emphasise that this form of selection is not mechanical but psychical. If Sexual Selection plays the great part in organic Evolution which Darwin, Weismann and many other great biologists assign to it, we can only conclude that to that extent at any rate the motive force in Evolution is psychical and not mechanical.

I would go further and, in opposition to current views, I would contend that even Natural Selection, in so far as it is a struggle among animal and plant forms and not merely the pressure or agelong effect of the inorganic environment, is fundamentally psychical. The advance takes place generally because the more fit organism deliberately for its own purposes destroys the less fit organism. In its essence the organic struggle creates a psychical situation just as much as war among humans is a psychical situation. And it is only because we abstract from the situation its real character of animal struggle and view the total statistical effect of innumerable situations as itself a sort of personified operative force in the form of Natural Selection, that the appearance of a mechanical external factor is created, operating on Evolution from the outside and determining its course mechanically. Looked at as a whole, and at a distance in which all concrete cases are blurred in the general view, the struggle for existence among animals and plants seems to operate blindly and mechanically without any view to that improvement of species which results. But this general struggle is actually composed of infinite little concrete struggles in which the fit deliberately destroy the unfit. And the trend of the struggle is towards organic progress actually because of the character of the little concrete struggles. I do not mean to say that the striving, struggling individual in nature intends to improve its species. But it does fight for itself or its family or its tribe; and in so far as it is more "fit" than its beaten opponent it is in effect fighting the battle of organic progress. The psychical purposive character of the little concrete struggles imparts a psychological purposive character to the generalised factor of Natural Selection. In fact the current view of Natural Selection is a very striking illustration of the way in which a so-called mechanical force or cause is gratuitously constituted by abstraction and generalisation and statistical summation from elements which in their individual character and isolation are undoubtedly psychical and purposive. And this only shows how careful we must be to scrutinise concrete details and not to rest satisfied with large generalisations, if we would know what really happens in nature. Abstraction and generalisation, however useful and necessary for scientific purposes, do largely deprive real events of their true characters, which are vital to a correct understanding of reality.

To sum up, therefore: apart from the influence of the physical environment, the motive and directive forces of organic Descent in the form of Natural and Sexual Selection are psychical and not merely mechanical. And this result of the special Darwinian theory is therefore in complete accord with the more general considerations which we derived from the analysis of Evolution in general. Both in Evolution as a whole and in Darwin's more special theory of organic Descent, life and mind are no mere shadows or unreal accompaniments of some real mechanical process; they are there in their own right as true operative factors, and play a real and unmistakable part in determining both the advance and its specific direction. From the point of view of Evolution each of them must be looked upon as essentially a real *vera causa*. This does not affect what I have already said about the vagueness and unsatisfactoriness of their present concepts and the necessity of looking for more definite and adequate concepts. All I mean to say is that the things they stand for are real factors in nature and not mere words or appearances. In the sequel an effort will be made to give greater definiteness to these concepts, and to determine the nature and character of the activity of these factors. Here it must suffice to emphasise that the nature of Evolution has been obscured by mechanistic conceptions, and that erroneous views as to the character and operation of causation have contributed to this misunderstanding. And it may be useful, before concluding this introductory chapter, to add a few remarks on this subject, to which I have already briefly referred above.

The science of the nineteenth century was like its philosophy, its morals and its civilisation in general, distinguished by a certain hardness, primness and precise limitation and demarcation of ideas. Vagueness, indefinite and blurred outlines, anything savouring of mysticism, was abhorrent to that great age of limited exactitude. The rigid categories of physics were applied to the indefinite and hazy phenomena of life and mind. Concepts were in logic as well as in science narrowed down to their most luminous points, and the rest of their contents treated as non-existent. Situations were not envisaged as a whole of clear and vague obscure elements alike, but were analysed merely into their clear, outstanding, luminous points. A "cause," for instance, was not taken as a whole situation which at a certain stage insensibly passes into another situation, called the effect. No, the most outstanding feature in the first situation was isolated and abstracted and treated as the cause of the most outstanding and striking feature of the next situation, which was called the effect. Everything between this cause and this effect was blotted out, and the two sharp ideas or rather situations of cause and effect were made to confront each other in every case of causation like two opposing forces. This logical precision immediately had the effect of making it impossible to understand how the one passed into the other in actual causation. The efficient activity, which had of old been construed on the analogy of our muscular activity, was therefore resorted to in order to supply the explanation. As the muscular movement produces external action, so material cause was supposed to produce a material effect. Even then the mind found it difficult to realise the passage from the one to the other. Every causation seemed to imply some action at a distance, unless cause and effect were in absolute contact. But we know that there is no such thing as absolute contact even in the elements of the most closely packed situation. Hence causation of this rigid type really became unintelligible. Not even the old fiction of an ether which embraced all material things, and as a vehicle made transmission of influence from one to the other possible, seemed able to overcome the contradictions into which thought had landed itself through its hard and narrow concepts of cause and effect. And in fact there is no way out of the *impasse* but by retracing our steps and recognising that these concepts are partial and misleading abstractions. We have to return to the fluidity and plasticity of nature and experience in order to find the concepts of reality. When we do this we find that round every luminous point in experience there is a gradual shading off into haziness and obscurity. A " concept " is not merely its clear luminous centre, but embraces a surrounding sphere of meaning or influence of smaller or larger dimensions, in which the luminosity tails off and grows fainter until it disappears. Similarly a " thing " is not merely that which presents itself as such in clearest definite outline, but this central area is surrounded by a zone of intuitions and influences which shades off into the region of the indefinite. The hard abrupt contours of our ordinary conceptional system do not apply to reality and make reality inexplicable, not only in the case of causation, but in all cases of relations between things, qualities, and ideas. Conceive of a cause as a centre with a zone of activity or influence surrounding it and shading gradually off into indefiniteness. Next conceive of an effect as similarly surrounded. It is easy in that way to understand their interaction, and to see that cause and effect are interlocked, and embrace and influence each other through the interpenetration of their two fields. In fact the conception of Fields of force which has become customary in Electro-Magnetism is only a special case of phenomenon which is quite universal in the realms of thought and reality alike. Every " thing " has its field, like itself, only more attenuated; every concept has likewise its field. It is in these fields and these fields only that things really happen. It is the intermingling of fields which is creative or causal in nature as well as in life. The hard secluded concrete thing or concept is barren, and but for its field it could never come into real contact or into active or creative relations with any other thing or concept. Things, ideas, animals, plants, persons: all these, like physical forces, have their fields, and but for their fields they would be unintelligible, their activities would be impossible, and their relations barren and sterile. The abstract intelligence, in isolating things or ideas, and constituting them apart from their fields, and treating the latter as non-existent, has made the real world of matter and of life quite unintelligible and inexplicable. The world is thus in abstraction constituted of entities which are absolutely discontinuous, with nothing between them to bridge the impassable gulfs, little or great, which separate them from each other. The world becomes to us a mere collection of *disjecta membra,* drained of all union or mutual relations, dead, barren, inactive, unintelligible. And in order once more to bring relations into this scrap heap of disconnected entities, the mind has to conjure up spirits, influences, forces and what not from the vasty deep of its own imagination. And all this is due to the initial mistake of enclosing things or ideas or persons in hard contours which are purely artificial and are not in accordance with the natural shading-off continuities which are or should be well known to science and philosophy alike. One of the most salutary reforms in thought which could be effected would be for people to accustom themselves to the ideal of fields, and to look upon every concrete thing or person or even abstract idea as merely a centre, surrounded by zones or aurae or spheres of the same nature as the centre, only more attenuated and shading off into indefiniteness.

There is one more remark I wish to make in regard to the activity of the abstract intelligence in construing our actual experience. I have already shown how in a special case this abstract activity has converted the psychical factor of Natural Selection into the semblance of a mechanical force. The risk of error is, however, much greater than that particular instance may serve to indicate. One may say that the analytical character of thought has a far-reaching effect in obscuring the nature of reality, which has to be carefully guarded against. In order to understand and explore any concrete situation, we analyse it into its factors or elements, whose separate operation and effects are then studied, in isolation so to say. This procedure is not only quite legitimate, but the only one possible, if we wish to understand and investigate the complex groupings of nature. It is the analytical method which science has applied with such outstanding success; and but for this analysis of a complex phenomenon or situation into its separate elements and the study of these in isolation, it is fair to assume that very little progress would have been possible in the understanding of Nature with all her obscure processes. When the isolated elements or factors of the complex situation have been separately studied, they are recombined in order to reconstitute the original situation. Two sources of error here become possible. In the first place, in the original analysis something may have escaped, so that in the reconstruction we have no longer all the original elements present, but something less. I have already shown how " fields " escape in the idea of things and even in concepts. The same happens in regard to the elements into which a situation is analysed. And it is certain that in every case of analysis and reconstitution of a situation something escapes, which makes the artificial situation as reconstructed different from the original situation which was to be explored and explained. An element of more or less error has entered. This may be called the error of analysis.

In the second place, we are apt after the analysis and investigation of the isolated elements or factors to look upon them as the natural factors of the situation, and upon the situation itself as a sort of result brought about by them. The analytical elements thus become the real operative entities, while the situation or phenomenon to be explained becomes their product or resultant. As a matter of fact, just the opposite is the case. We started in nature with the complex situation or sensible phenomenon as the reality to be explained. The analytical elements or factors were merely the result of analysis, and might even be merely abstractions. But because they are simpler and admit of closer scrutiny and experiment, we have come to look upon them as real or constitutive, and upon the situation from which they were abstracted or analysed as artificial or constituted. Thus it has come about that in physical science, for instance, the elements of matter or force into which bodies have been analysed have tended to become the reals. Thus scientific entities like electrons and protons, and the physical energies or forces which they represent, are taken to be the real entities in nature, and sensible matter or bodies as something derivative and merely resulting from their activities. The abstract thus becomes the real, the concrete is relegated to a secondary position. This inversion of reality is very much the same procedure as was condemned in the case of the scholastic and other philosophers who attributed reality to universals instead of to concrete particulars. This may be called the error of abstraction or generalisation. Against both these forms of error we have to guard, if we wish faithfully to interpret Nature as we experience her.

Our object in studying and interpreting Nature is to be

faithful to our experience of her. We do not want to recreate Nature in our own image, and as far as possible we wish to eliminate errors of observation or construction which are due to us as observers. We do not wish to spread Nature on a sort of Procrustes bed of our concepts and cut off here and there what appears surplus or unnecessary or even nonexistent to our subjective standards. Our experience is largely fluid and plastic, with little that is rigid and with much that is indefinite about it. We should as far as possible withstand the temptation to pour this plastic experience into the moulds of our hard and narrow preconceived notions, and even at the risk of failing to explain all that we experience we should be modest and loyal in the handling of that experience. In that way a good deal of what we have hitherto felt certain may once more become uncertain; the solid and recognised landmarks may once more become blurred or shifting; the stable results of the nineteenth-century science may once more become unstable and uncertain. But the way will be open for the truer constructions of the future, and the foundations of our future science will be more deeply and securely laid.

In the following chapters a modest effort will be made to apply the above ideas and principles to a new interpretation of Nature, including, as it does, Matter, Life and Mind. Matter, Life and Mind, so far from being discontinuous and disparate, will appear as a more or less connected progressive series of the same great Process. And this Process will be shown to underlie and explain the characters of all three, and to give to Evolution, both inorganic and organic, a fundamental continuity which it does not seem to possess according to current scientific and philosophical ideas.

CHAPTER II

THE REFORMED CONCEPTS OF SPACE AND TIME

Summary.—It is not only in organic Evolution that the old fixed concepts and counters of thought are breaking down. Recent advances in physical science have extended the revolution to the domain of the inorganic; the fixity of the atom has followed that of species into the limbo of the obsolete. In many directions new concepts, more in harmony with the fluid creative process of nature, are called for.

We begin with the new concepts of Space and Time, which in the system of Relativity are taking the place of the old Newtonian concepts still commonly accepted. The new ideas of Space and Time arose from researches in the higher mathematics and physics, and were primarily concerned with the relative character of all actual motion in the universe, and the mathematical and physical consequences of this relativity. Thus according to the mathematical physicists to a moving observer a moving body appears to contract or to be shorter than it would be to a stationary observer, and the faster either of them moves the greater the contraction becomes. Time varies similarly, but in the opposite direction; while the space of the moving body appears to contract, its time appears to expand, so that it takes a longer time to pass a point than it would do if viewed by a stationary observer. This joint and inseparable variation of Space and Time was not only most important in itself, but led directly to the revolutionary conception that neither of them existed independently, but that together they form the Space–Time medium of the real physical world. From this point of view bodies and things as merely spatial are not real but abstractions, while events, which involve both Space and Time, Action in Space–Time, are real and form the units of reality. The deposition of the old Space and Time and their replacement by Space–Time have been tested in the most searching way both in the immense world of astronomy and in the most minute world of the atom, and in both cases the new concepts have been found to work satisfactorily.

The variation of Space and Time has led to the further conclusion that in a world of relative motion such as ours all standards of measurement and all clocks of Time are themselves variable and

CHAP. II CONCEPTS OF SPACE AND TIME 23

give no constant results. Applying the conclusion to gravitation and the rotational movements of the universe, we find that the Space–Time medium of the universe is curved and warped and not of the homogeneous character which was attributed to Space and Time according to the old ideas. In all gravitational fields events happen in curves and follow the fundamental curves of the Space–Time universe. The result is that the entire universe acquires a definite structural character, and is not a diffuse homogeneity as was formerly supposed. According to the new Space–Time concept, structure, definite organised structure, becomes the essential characteristic of the physical universe, and this structural character accounts for many hitherto inexplicable phenomena.

IN the preceding chapter I have tried to explain how the acceptance of the theory of Evolution must inevitably and profoundly affect our views as regards the nature of matter. In this chapter I proceed to inquire what bearing recent far-reaching physical researches and speculations have on this position. Our problem is to break away from the hard and narrow conceptions of the Victorian age, to see Nature once more in her fluid and creative plasticity, and to formulate our conceptions afresh from this deeper point of view. A great change has come over our views of Nature, a change great enough in the end to amount to one of the fundamental revolutions in human thought. But we are still in the process of that change, and it is therefore difficult for us to realise its full significance. Three dates stand out in bold relief as inaugurating that change: 1859, when Darwin's *Origin of Species* was published; 1896, when Becquerel discovered Radioactivity; and 1915, when Einstein published his *General Theory of Relativity*. Round these three great events other discoveries of profound interest have taken and are still taking place; and in the result our entire viewpoints and standpoints as regards Nature and reality are undergoing a fundamental change which must in the end affect every province of human thought and conduct. The fixity of organic species is gone; the fixity of inorganic elements is going. The position is once more becoming fluid, the old rigid order is visibly dissolving, the fixed landmarks and beacon-points by which former generations steered their course in science are becoming submerged. And the task awaits the future out of this fluid situation and these instabilities once more to build a stable world of ideas, which will be in closer harmony with the reality around us and within us. One of the aspects of Darwin's Theory has already briefly engaged our attention in the last chapter, and other aspects of it will be considered in Chapter VIII. In the present chapter reference must be made to Einstein's *General Theory of Relativity* and the bearing it has on our ideas of space and time as the framework in which events are located, and the medium in which Evolution takes place. The resulting view of the universe as structural, and of the element of structure as fundamental to the universe and all its forms, is important for the subject and the argument of this work.

People become frightened when they are invited to consider Einstein's theory. Its refined abstractions, its abstruse mathematical form, its complete novelty and reversal of ordinary common-sense view-points make it a terror to the uninitiated. And yet I believe the Einstein view-point can be quite simply and intelligibly put. Indeed it must be so put if it is ever to become part and parcel of ordinary educated thought. We must distinguish between the simple and clear view-point itself, and the recondite mathematical processes by which it was reached, and the technical mathematical form in which it is expressed, and from which for all ordinary purposes it can be separated. The understanding and appreciation of the Relativity view-point are not dependent on a knowledge of the process by which Einstein reached that view-point. The result is quite distinct from the process. It is like groping our way through a long, dark, rough tunnel, and at the end emerging into the clear daylight beyond: it is not necessary for the appreciation of the new view that one should plunge back into the dark tunnel. Besides, I must frankly state my own opinion that the Einstein theory, as distinguished from the broad view-point attained, has not yet found its final expression. All great truths are in their essence simple; and the absence of simplicity of statement only shows that the ultimate form has not yet been reached. The day may yet come when the ten recondite Einstein equations of gravitation may appear as but the scaffolding of the simpler structure yet to arise, the naturalness and inevitableness of which will be as evident to every educated person as the heliocentric conception of Copernicus has become.

The Einstein theory arose originally from mathematics, and a brief reference to this mathematical origin will be useful. Galileo and Newton were the fathers of the modern classical mechanics; they (and especially Newton) formulated the laws of moving bodies in an exact mathematical science. Now the germ of the new Relativity mechanics is the almost obvious fact that the motion of a body is never absolute, but is always relative to some other body or point. If this body or point of reference is stationary, Newton's laws of motion apply completely, and the geometry of Euclid also applies, so that the movements of bodies can be represented by geometrical figures. Such bodies are said to move in Euclidean space, which is the same and homogeneous all through and in all directions. Now since Newton's time a great deal of attention has been given to the case where the body of reference or the observer is not stationary but is also in motion. This case is important, because it is actually that of all bodies in our universe, in which all observers or points of reference are themselves in motion. A point on the earth, for instance, rotates with a certain velocity round the centre of the earth, while the earth again rotates with another velocity round the sun. The sun itself is not stationary but moving with reference to some star, which is itself in motion with reference to some other moving centre of reference. It is this case of the moving observer or point of reference with which Einstein's theory deals, and it is therefore clear that this theory faces the problem of motion as it actually exists in the universe. The impression of rest or stationariness to us as observers in the universe is a mere illusion, and the great service of Einstein was to explore this illusion and to show in exact mathematical form to what extent it affects our vision and judgment of movement in the world. Let us therefore take the case of moving observers. Now when a moving object (say a train in motion) is viewed by an observer in motion (say an observer in a motor-car moving on a road parallel to the train), certain curious results have been worked out by the mathematicians, of which the following are two important samples:

(A) The train appears a little shorter than it would to a stationary observer.

(B) The time taken by the train to pass a point appears a little longer (or the train appears to move somewhat more slowly) than it would to a stationary observer.

In other words, to a moving observer the length or the space occupied by a moving body is smaller in the direction of its motion than it would appear to a stationary observer; and similarly the time taken by the observed body to pass a point will be longer. And the faster the observer or observed body moves, the more the space and time of the observed body will vary for him, compared to what they would do if he were at rest. These two variations of space and time are joint variations, happening simultaneously but in an opposite direction, the one becoming less in proportion as the other becomes more to the moving observer. Space contracts and time expands in inverse proportions according to the rate of motion of a moving body of reference or a moving observer. One may generalise this result and say that so long as several observers move at different rates but uniformly and in straight lines with regard to each other, the velocity or speed of the moving body which they are observing appears the same to all of them, as the proportional co-variations of their respective spaces and times cancel each other out, so to say. This is a popular way of stating the main principle of Einstein's *Special Theory of Relativity*, first published in 1905, in rigorous mathematical form. It explained the fact, which had been repeatedly confirmed by the most accurate experiments, that the velocity of light is always the same, whatever the velocity of its source, and however great may be the difference in velocities of the moving observers who are trying to measure it. An observer moving away from a flash of light at a rate which is half that of light will see the flash at the same time as a stationary observer, and not later as one might suppose. The reason is that the time and space measures of the moving observer have changed jointly so as to neutralise the results which his motion might have on his observation.

The two salient facts to bear in mind as a result of the above are: that to moving observers clocks and standards of measurement in motion are no longer absolute but vary according to the rates of motion of these observers or of the clocks or standards, and that there is this curious joint and opposite variation of the space and time measures of moving observers or bodies. In fact separately Space and Time must be mere abstractions, as in all actual movements they are always found in inseparable conjoint action.

From this co-variation of Space and Time it is but a step to Minkowski's great idea, first formulated in 1908, that in natural events Space and Time are not independent factors, and that in the mathematical representation of events the correct way is to introduce time as a fourth dimension, not of space, but of the Space–Time continuum in which events really take place. Time is, of course, in many ways unlike space and is not another dimension of it, but this inseparable co-variation in all events which happen in nature makes it both feasible and proper that we should substitute the real Space–Time continuum of events for the old abstract three-dimensional space of bodies or points in space. In passing it may here be pointed out that the old notion of the separate reality of space and of time involved both the errors of analysis and of abstraction to which attention was drawn in the last chapter, and Minkowski's brilliant idea has simply brought us back to the natural fact as it occurs in experience, where nothing ever happens in space alone or in time alone, but always in both together, and where objects are not observed by themselves, but always as elements or items in the stream of perceived events. Nay more, it can be easily shown that the very ideas of Space and Time interpenetrate each other and are dependent on each other. Succession or the time-series, and co-existence or the space-series, are necessary to each other and would not be even intelligible apart from each other. For the succession (time) would perish at each step and would not even form a series, unless it had enduringness or co-existence (space). And similarly the co-existence (space) would stop at its first step and would not be spread out or extended unless it had also succession (time).

To Einstein this concept of a Space–Time continuum proved most welcome and fruitful, and he proceeded to apply it to the explanation of all movements in the universe, not only to uniform and rectilinear motions which take place in uniform Euclidean space, but also to rotating and accelerated motions which take place in a gravitational field of a non-Euclidean character.

His first step was to illustrate, by purely theoretical considerations, the fact that a body under the influence of a constant force, and therefore moving with a constantly increasing acceleration, would to an observer situated on it behave in exactly the same way as a body acted on by gravitation. Thus suppose a man enclosed in a cage so that he cannot observe any other body and cannot notice his own motion. And suppose this cage suspended in distant space where there is no gravitation. And suppose further that this cage is drawn upward with a constant force, so that it moves faster and faster with a constantly increasing acceleration. In truth the case is therefore one of acceleration. But by the enclosed observer this motion and acceleration of the cage and himself will not be noticed. He would only feel like being pulled down by his own weight. If he loses any object from his hand, it will fall to the bottom of the cage like a stone dropped on the earth. What is more, the rate at which the object drops is the same, *whatever its figure or size or amount of material.* The fact italicised is distinctive of gravitation. In other words, the man in the cage will think that he and the cage and the object therein are all acted upon by gravitation. What is really due to acceleration appears to be a case purely and simply of gravitation. Thus we see acceleration and gravitation are really the same phenomena and only different in appearance to observers. Acceleration and gravitation are, in fact, equivalent expressions. Einstein's closed cage may yet become as historic as Newton's falling apple.

Now take rotation, which is simply a special case of acceleration. And let us imagine an observer situated on a rotating or revolving plane circular disc and proceeding to measure the area of the disc and the rate at which it is revolving. He has two identical clocks, one of which he puts near the centre of the disc and the other near the circumference in order to take some time measurements. When he proceeds to take the time of the clock near the centre he finds that it moves more slowly than when he proceeds to read the clock placed near the circumference. We have already seen why this is so. The motion of the disc at the centre is nil, and its motion at the circumference quite marked, and the times of identical clocks at these two points will therefore vary to the moving observer. And similarly the rate of any identical clock will vary according to the distance of its position on the surface of the disc from the centre, as the motions of all points on the disc will differ according to their distance from the centre. He then proceeds to apply identical measuring rods and finds the same continual variation. He finds that the identical measuring rods vary in length according to their position on the disc; one placed on the circumference is shorter than one placed near the centre. And the differing lengths of the rods will measure up different spaces. The observer will become utterly confused, and will finally conclude that the spaces on the disc are not the same everywhere and in all directions, but appear to vary in all directions and to be twisted, warped, and curved. Or, as we would say, the space of the disc is not straight-line homogeneous uniform Euclidean space, but curves; and these curves are nothing but the actual orientation or direction of events in the Space–Time framework of the universe.

What is or would be the situation beyond the material universe and its vast fields? There we pass beyond the bounds of gravitation, where there is neither rotation nor acceleration, where " bodies " (if such astral abstractions could be imagined) persist in their state of rest or of uniform motion in a straight line according to Newton's First Law. There Space–Time, if it could be imagined to exist, would not be curved, but would be homogeneous and continuous, and would be exactly the form of empty nothingness. In fact, homogeneous Euclidean Space–Time beyond all real fields is simply a limiting conception of thought and would correspond to nothing that we have any knowledge of in our universe.

It may be interesting, in conclusion, to point out the difference between this conception of Space–Time in the Relativity Theory, and the conceptions of Space and Time formulated by Newton and Kant. For Newton both Space and Time were absolutes; that is to say, were real invariable permanent characters of things and events. They were each homogeneous and continuous and therefore adequately expressible by the geometry of Euclid. There was nothing subjective about them. For Kant, who in other respects profoundly admired the Newtonian system, the great problem of knowledge was how to determine the relative contributions made to our knowledge of the world by the subjective and objective factors respectively, and especially how much and what the mind brought into the common pool of knowledge and experience. His answer in effect was that the action of the mind was creative in experience and that it contributed to our knowledge—(*a*) the elements of Space and Time which are nothing but the mind's own sensuous forms of intuition or perception imposed on the materials of sense and experience, and (*b*) the general conceptual system of knowledge which follows from the categories of the Understanding, and (*c*) certain ultimate regulative principles of the human Reason. According to this view, therefore, Space and Time are nothing but the necessary forms of man's sensuous perception; they do not exist in external reality, but are imposed by the mind on all objects of sense. While accepting the homogeneous universal Euclidean characters which Newton ascribed to Space and Time, Kant denied that they were characters of things or events. If these characters belonged to things, Kant failed to understand how the *a priori* synthetic character of mathematical knowledge was possible, and he could only explain this fact by making the sensuous form of things a subjective contribution of the mind itself. The universal forms of Space and Time in knowledge were due, not to the things or the world to which they seemed to belong, but purely and simply to the perceiving mind which invested all things with them.

In contradistinction to both these theories, Space and Time in the theory of Relativity as conjoint co-ordinate forms belong both to the mind and to things; and the whole effort of Einstein was to separate the subjective appearance from the objective reality, to separate the relative, variable and disturbing contribution made by the observing mind from the real permanent Space–Time factor which is inherent in the physical universe. If the confirmation of theory by facts means anything it must be admitted that Einstein has been singularly successful in his analysis and evaluation of these two subjective and objective aspects of the Space–Time concept. That Space and Time were not, on the one hand, merely subjective conditions of experience as Kant held, nor, on the other, merely objectively given elements for experience as Newton held, but that they were both subjective and objective contributions to experience, might have been the discovery of a sound psychology or epistemology. But that these two factors of Space and Time have been fused into one synthesis, from which both the subjective and objective elements have been properly sorted out and isolated and valued and rigorously determined, is an achievement of the most outstanding importance not only for science but also for philosophy. It is unnecessary to point out that in the *Theory of Relativity* Space and Time are not metaphysical conceptions or forms. The infinite homogeneous Space and Time which to Kant were mental presuppositions and preconditions of all sensuous experience, and to Leibniz the pre-established permanent universal order of co-existence and succession among things, are to Einstein mere limiting pseudo-concepts, metaphysical abstractions without relation to our real experience. In our experience Space and Time are given elements just as colour, weight, and the rest. The task of science is to co-ordinate these elements in an intelligible form, and in doing so Einstein has simply explored them as if they were real physical experience like the rest. The result is the elimination of certain historic errors from the concepts of Space and Time, and the determination of their physical qualities in line with the rest of our physical experience and concepts. The Space–Time continuum, instead of being a vague, homogeneous, formless, metaphysical concept, becomes a part of physical reality, becomes the " field " of the material world, with a definite structure of its own. Structure, real differentiated structure, becomes the inmost form of the real Space–Time world. The close bearing of this on the main argument of this work will appear from the following chapters.

CHAPTER III

THE REFORMED CONCEPT OF MATTER

Summary.—Coming now from the Space–Time continuum to Matter we find the feature of structure much more conspicuous and important. The physical and chemical constitution of matter is almost entirely a matter of structure. Chemistry has traced matter to its ultimate units or atoms, and to the combination of these into molecules and substances according to structural schemes dependent on the placing and spacing of the different units in the various chemical combinations. The New Physics has carried the process a step further back by analysing atoms into their constituent electrons and protons, or units of negative and positive electricity. These units are so arranged structurally as to approximate to the form of more or less complicated solar systems, with central positive nuclei and revolving planetary electrons. The explanation of the physical and chemical properties of matter has been traced to the structural arrangements in these atomic systems and the number and changes in position of their various units. Matter is thus a structure of energy units revolving with immense velocities in Space–Time, and the various elements arise from the number and arrangement of the units in an atom; as these can be varied, the transmutation of elements becomes possible, as in Radioactivity. The peculiar serial character of the Periodic Table of the elements is thus due to the number of units and their architecture in the atoms. Atomic Weights and Atomic Numbers reflect this inner arithmetical character of the atoms.

The states of matter, as gaseous, liquid or solid, are also the results of the residual surface forces in atoms and molecules, due to their inner structure. Crystal structure is another result of inner atomic structure. But perhaps the most remarkable state of matter is a combination of the other states; this is called the colloid state, in which very minute particles of one material are dispersed throughout another. This colloid state is much more universal than commonly thought, and is specially important because the protoplasm of cells is organised in this state. The minuteness of the dispersed particles means the exposure of a maximum surface area compared to their mass. These surfaces bring into play the surface forces and show peculiar affinities or selective properties of various kinds, and in this way certain chemical and physical reactions are facilitated at these surfaces, which make them useful in the industries as well as in the processes of organic life. In fact, some reactions in the colloid state approximate strangely to the biological type.

From the above analysis of the structural energetic constitution of matter certain conclusions can be drawn which very much narrow the gulf between matter and life.

In the first place, the old view of matter as inert and passive disappears completely. Matter like life is intensely active, indeed is Action in the technical physical sense; the difference is not between deadness and activity, but between two different kinds of activity. Through their common activities the fields of matter and life thus overlap and intermingle, and absolute separation disappears.

In the second place, Radioactivity in matter plays a somewhat analogous rôle to Organic Descent in life. Both render fluid the old fixed entities and forms; although the difference between them must not be minimised. Especially must it be recognised that Radioactivity is regressive, while Organic Descent is progressive. But this may be due to the extreme age of matter as compared with the youth of life in the history of the earth.

In the third place, the Periodic Table of Chemistry has a distinct resemblance to the Systems of Botany and Zoology; the concepts of families, genera and species could be applied to both. This shows that the characters of activity, plasticity and probably of development and genetic relationships apply to both the organic and inorganic domains.

In the fourth place, the structural character of matter indicates that it is also creative, not of its own stuff, but of the forms, arrangements and patterns which constitute all its value in the physical sphere. Just as life and mind are creative of values through the selective combinations and forms which they bring about, so matter also, instead of being dispersive, diffusive and structureless, effects through its inner activities and forces structural groupings and combinations which are valuable, not merely to humans, but in the order of the universe. But for its dynamic structural creative character matter could not have been the mother of the universe.

In the fifth place, matter in its colloid state in protoplasm discloses properties and manufactures substances, such as chlorophyll and hæmoglobin, which are necessary for the functions of life, and which go far toward bridging the great gap between the two. In its colloid state we thus see matter reaching up to the very threshold of life. A great leap may have taken place across what remained as a gap. A great "mutation" may have occurred. But as life probably began on a much lower level than the lowest forms we know to-day, the mutation may after all not have been so great. In any case a close scrutiny of the nature of matter, as revealed by the New Physics, and especially colloid chemistry, brings it very near to the concept of life.

LET us now proceed to consider how recent advances in our knowledge of the constitution of matter have emphasised the importance of this same feature of structure in the physical universe. Chemistry had for a century been exploring with great success the structure and constitution of matter, but the New Physics of Radioactivity has during the last twenty years proved a most powerful aid to Chemistry and led to discoveries which are little short of revolutionary. To Chemistry was due the analysis of matter into a certain number of elements, each with its own physical and chemical properties; the discovery of the atom as the ultimate unit of each element of matter; the union of atoms of each element into simple molecules of that element, and the union of atoms of various elements into compound molecules. The combinations of elements in definite quantitative proportions was explained as the union of one or more of the atoms of these elements with each other. From this it might be inferred that the combinations of Chemistry were like the combinations of Arithmetic, and that the whole numbers of Arithmetic might properly represent the atoms of Chemistry and their combinations into compounds. This inference has, however, only been actually verified by the recent physical discoveries. It was not only the fact of numerical or quantitative structure that was important to Chemistry; the spatial or positional structure of matter, the order of placing and spacing of the atoms in the chemical substance, the architecture of matter became almost equally important, and in many cases the properties of a substance could only be explained on the basis of its real or supposed inner structure and configuration. Thus molecules of carbon could be either coal or graphite or diamond, and this great dissimilarity in the same chemical substance was explained as the result of the difference of structure in the placing and spacing of the atoms in the carbon molecule. Sulphur and many other elements show a similar polymorphous or, as it is called, allotropic character. It was, however, when chemists had to explain the different characters of quite distinct chemical compounds, which yet had the same chemical composition, that the importance of "structure" and constitution became most highly accentuated. Such compounds are called isomers. So important is structure to matter that without it one may safely say that organic chemistry becomes unintelligible. The more complex the composition of substances (as in organic chemistry), the larger the number of permutations and combinations that are possible in the relative positions and placings of atoms or groups of atoms in the make-up of matter, the more important does the phenomenon of isomerism become, and the greater is the part played by structure and configuration in the building up of matter. The chemical formula is no longer sufficient, it is a mere abstract notational shorthand which may be thoroughly misleading in the absence of a diagrammatic representation of the constitution or structure of the compound substance. The crystal forms of solids illustrate not only the structural character of chemical substances, but also the invariable way in which the same substance follows the same pattern of structure. To Chemistry structure, or the proper representation of relative positions of atoms or their groups in the three dimensions of space, has become indispensable. And the New Physics has now gone a step further and shown that this minute structure of the chemical atom and compound is not static in space, but dynamic and intensely active in that Space–Time continuum which we have already found dominant in the relations of astronomical bodies and events. Space–Time prevails at both ends of physical infinity and everywhere between.

To Chemistry the atom was a hard indivisible unit, the constitution of which (if there was any) could not be known; nor could it explain chemical affinity or why atoms combined into molecules; nor could it explain the strange serial character of the Periodic Law in reference to the atomic weights and the properties of atoms. These triumphs were reserved for the New Physics, and they have traced structure back into the innermost recesses of the atom. The discovery of Radioactivity by Becquerel in 1896 at Paris was the first indication that the atom was not indivisible and could break up spontaneously in nature. The discovery in the previous year of the X-rays by Röntgen for the first time revealed the existence of invisible rays whose wave-length was as small as atoms, and the elaboration of the spectrum of these rays has provided an instrument of incredible power and accuracy in the investigation of the almost infinitesimally small phenomena of atomic structure. Then followed in 1899 the isolation by Sir J. J. Thomson of the ultimate unit of negative electricity in the electron; and in the following year Max Planck of Berlin University discovered what came to be known as the *quantum*, the unit of radiant action emitted by all radiant bodies or even dark bodies. The application of these new ideas and means of investigation by a number of brilliant researchers has led to the elucidation of the nature and constitution of the atom of matter in the theory which is specially associated with the names of Sir Ernest Rutherford and Professor Niels Bohr. Without entering into details which do not concern us, and simply to illustrate the element of "structure" in the atom with which we are dealing, I shall summarise the salient points in this theory. According to it, an atom is an electrical constellation somewhat like our gravitational solar system; the centre of the system being a minute very massive nucleus positively electrified, round which revolve equally minute electrons or negative particles of very small mass—so small that in the Hydrogen atom, for instance, the nucleus has 1835 times the mass of the electron. The electrons revolve at various rates in their different orbits, all of which can be measured through their X-ray spectrum; and an electron can suddenly and all at once jump from one orbit to another, increasing its orbit when it receives one or more quanta of radiation from some outside body, or decreasing its orbit and taking one nearer the central nucleus, and in the act of doing so releasing one or more quanta of radiation. It is these quanta of radiation, released when the electron jumps to a narrower orbit in the atom, that account for the light which comes from the sun and the stars, and in fact all radiant bodies; and it is the definite quanta of radiation so emitted which account for the peculiar spectrum of the elements in the spectroscope. Why atomic light should be emitted in these definite amounts or quanta is not yet known, but it is known that the quanta follow a scale somewhat similar to the notes in music, and we may therefore think of light as the music of the spheres, in which the total harmony or light effect is made up of definite discontinuous notes instead of continuous variations of light. The wonderful thing is that in regard to all these matters we have the most minute and accurate knowledge: the amount of a quantum; the mass, velocity and orbits of an electron; the mass and velocity of rotation of a nucleus, and the total sphere of an atom, with its small nucleus and electrons and vast empty spaces, comparable to the empty spaces in our solar system. The electron is by now very well known, and indeed all electric currents are nothing but streams of free electrons. But of the corresponding positive unit which is called a proton next to nothing is known, as the proton has never yet been isolated. Now the nucleus of an atom may be simple or complex; it may be a proton, as in the case of the Hydrogen atom, or it may consist of several protons, some of which, again, may be neutralised by closely associated electrons, and some remain unneutralised, so that the nucleus as a whole always remains positively electrified. In the Hydrogen atom there is one proton in the nucleus, and hence there is one electron revolving round it. In the Helium atom, again, there is a nucleus of four protons, two of which have electrons in association with them, and two not; the nucleus, therefore, has two positive units, to which correspond two electrons which revolve round the nucleus in the atom. The combination of four protons and two electrons in the Helium nucleus appears to persist in other nuclei, so that the nuclei of the other elements appear to be a combination of simple

III REFORMED CONCEPT OF MATTER

(Hydrogen) and complex (Helium) nuclei. The number of revolving electrons in an element always corresponds to the number of unsatisfied positive units in the nucleus, which is called the atomic number of the element, and which is always an integer; and thus the atomic numbers of the elements run from 1 in the case of Hydrogen to 2 in the case of Helium, 3 in the case of Carbon, and so on to 92 in the case of Uranium, the heaviest of all the known elements. Of these 92 possible elements, 5 or 6 have not yet been isolated, although their atomic numbers and weights and positions in the Periodic Table and their approximate properties are known.[1] Atomic weights are in every case integers, the apparent exceptions being cases where we have to do with isotopes or elements of which the atoms are not all identically the same but slightly different in their electron contents. Thus the New Physics has incidentally explained the mystery of the Periodic Table.

The mystery of the atom has now largely narrowed down to the nucleus, which consists of an inner revolving system of protons of which comparatively little is known except that they rotate round their centre with an enormous velocity—probably not much less than that of light—and that the quantum law as well as the mass law of Relativity holds with regard to them. As space or volume contracts with velocity in Space–Time, the mass of the nucleus increases with high speeds out of all proportion to its size, and the positive nucleus of the atom is therefore its virtual mass, the rest of the atom being either empty space or very light insubstantial electrons. Owing to its massiveness the nucleus of protons is therefore coming to be identified with matter, as if matter were ultimately only high-speed densely massed positive electricity. The proton may thus yet prove to be the fundamental unit of matter. The significance of this view is that it reduces matter simply to a form of energy,

---

[1] If the claims to the recent discoveries of Hafnium, Masurium, and Rhenium are allowed only a couple of further elements await discovery.

or rather Action, and therefore still further simplifies the scheme of the universe.

There is another fact which shows the intimate relation between energy on the one hand and structure or mass on the other. The mass or atomic weight of the free Hydrogen atom has been determined as 1.0077. In the Helium nucleus, as we have seen, there are four protons or Hydrogen nuclei, but here their mass only appears as one. In other words, the free Hydrogen atoms or protons (they are practically identical as regards mass) suffer a diminution of mass when they are concentrated into the Helium nucleus, as if in this nucleus, which is itself an inner constellation system, the protons and electrons are so close as to jam each other, and therefore move more slowly and thereby decrease their mass or matter. When the Helium nucleus is again split up into Hydrogen protons, this loss of mass would be recoverable in the form of energy, which, small as it is in the Helium nucleus, must be enormous in the world, as in all matter the nuclei are composed either of Hydrogen protons or Helium protons (their compressed form) or both. Should this energy ever become economically available, the greatest potential source of energy in the universe will be opened up for the benefit of mankind.

This would involve the artificial breaking up of matter, and this is the phenomenon which we actually witness in a natural spontaneous form in Radioactivity. In Radioactivity the nuclei of the heavier elements (Uranium, Thorium, and Radium) spontaneously break up and eject Helium at an invariable slow rate, which is regular enough to be a geological clock, now being used as a measure to calculate the age of the oldest rock-formations of the earth.[1] Thus the Periodic Table shows that the expulsion of three Helium atoms from Uranium will convert it through Thorium into Radium; the expulsion of one more Helium atom will convert Radium into an

---

[1] The age of these oldest formations, the Algonkian mountains of Canada, has thus been calculated as 1400 million years. Thus on this basis we obtain the lowest limit for the age of the earth.

element called Radium Emanation; and so on until eight Helium atoms have been expelled, when Lead will be reached. If the process of expulsion could be continued, Mercury will next be reached, and next after that Gold. The alchemists were then not so far out when they guessed that Mercury could be transmuted into gold! Unfortunately (or rather I should say fortunately as a citizen of the greatest gold-producing country) this spontaneous break-up of matter has not yet been observed to proceed beyond lead. And the artificial break-up of matter in the laboratory has only just begun in the experiments of Sir Ernest Rutherford, who by bombarding Nitrogen gas with α particles from Radium C has succeeded in splitting the Nitrogen atom into Hydrogen atoms and a residual apparent combination of Helium nuclei which might result in Carbon according to the Periodic Table, but which is more likely to split up into Helium atoms. To what extent this artificial destruction of the elements is possible, and whether, if possible, it would be economically feasible, are questions for the future to answer.

We have seen that the positive charges of the nucleus have to be balanced by the corresponding number of negative electrons grouped in their orbits round the nucleus. On the number and grouping of these planetary electrons the external physical and chemical properties of the atom will depend. If the orbits followed impose a strain on the equilibrium of the atom, a quantum adjustment to a different orbit will be made. If the number of electrons and their orbit distributions produce complete equilibrium the atom will be very stable internally and inert or inactive externally; it will belong to one of the inert group (Helium, Argon, Neon, Krypton). On either side of this inert group of elements in the Periodic Table we find elements whose atoms have one electron too many or too few; in other words, they are not internally in equilibrium and have a negative or positive charge unsatisfied; they will therefore combine with any other element which has an opposite charge unsatisfied. At another remove from the inert elements in the Periodic Table we find elements with two negative or positive charges unsatisfied, which will again combine with another element which has two opposite charges unsatisfied. And so on to the elements which have three, four or five charges unsatisfied. In this way both chemical affinity and the valency (monovalency, divalency, etc.) of the elements are accounted for. In every case the external properties of the element are simply the expression of its internal structure and its condition of stable or unstable equilibrium in respect of its inner elements.

Not only the combination of atoms into molecules, but the formation of the most complex compounds rests on this condition of unstable equilibrium due to unsatisfied negative or positive charges in the combining elements. The compound, instead of being a single system of the solar type, is a far more complex affair, and represents the case where suns with their attendant planets again revolve round a greater central sun, or where several solar systems are linked together externally and not by a common centre. In either case the distribution and equilibrium of the moving internal electric units determine the structure of the substance as matter as well as its physical and chemical properties, while the movements of the substance as a whole and of its parts relatively to each other create the gravitational field or curved Space-Time system which forms the medium and the field of the substance. There is thus structure through and through, not only in matter or the energies which in their extreme concentration and velocity assume the massive form of matter, but also in the field which surrounds this matter.

The gaseous, liquid and solid forms of matter are also the result of this inner condition of electrical stability in the atom and molecule. If the positive and negative charges are quite equal and properly distributed the result, as we have seen, is an inert element. And this element will also be a gas, as the inner satisfaction of the charges and balance of the system will make it inactive or inert externally. All gases are states of matter where the inner balance of equilibrium in the atoms and molecules is such that there is no residual force to work externally; the atoms (in inert elements) and molecules (in others) therefore move freely and unhampered. If the inner balance of charges is not quite complete, there will be some external residual force as between the molecules, and the liquid state will result. If this inner satisfaction is lessened still further, the resultant external strain among the molecules will increase, they will attract each other still more strongly and tend to closer aggregation, and thus the solid form will appear. The negative or positive electrical condition of the gas, liquid or solid will be an index of the still unsatisfied charges residing in the substance in that state. The free and unhampered movement of atoms in an inert gaseous element and of molecules in other gases makes the question of the particular forms of such elements or gases immaterial; they have as gases no particular form. In the case of liquids, however, the resultant residual forces of the atoms and molecules will, as is the case in electrical bodies, act mostly at the surface, where the resulting force between the molecules of the surface layer, or the surface tension, as it is called, will give a particular form or shape to the liquid (as in a drop of water). The molecules inside a liquid appear to be stratified into layers loosely superimposed on each other. And in the case of solids the still larger residual force will result in arranging the molecules in a definite crystal structure on the pattern of a lattice, which is the special and specific form of solid chemical substances. Crystal structure is to solid compounds what the planetary structure is to the atom—not only a specific ordering of inner units, but the index and source of all external properties and activities. One of the most interesting recent discoveries is that in crystals there is a unit body or minute structure consisting of two or more molecules which is of atomic or radicle character in that it always acts as a unit in the upbuilding of the crystal.

Besides the gaseous, liquid, and solid phases of matter just discussed there is a fourth, to which in recent years an ever-increasing amount of attention has been and is being devoted. This is the colloid state, in which one substance is dispersed throughout another in very minute particles which are yet larger than molecules. Originally substances were divided into colloids and non-colloids; but more recently it has been shown that non-colloids (like mineral salts) can under certain conditions be reduced to the colloid state. And now this division has been abandoned, and the colloid state is recognised as a fourth form of material aggregation applying to substances generally under certain conditions. Much of the earth and the air exists in the colloid state; but the colloid state is specially important because it seems to be distinctive of all life-forms—the protoplasm of all organic cells being organised in the colloid state. The protoplasm of the cells contains solid substances in most minute form dispersed throughout its jelly-like fluid, and this colloid state seems to link the inorganic with the organic elements in the cell.

Owing to their minute size, particles in the colloid state expose the maximum surface area in comparison with their mass; and the colloid state in consequence brings into action the play of surface energies more than any other phase of matter. In all forms of aggregation the surface molecules of matter are specially orientated; the active sides of the molecules being turned inwards, and the outer surface thus consisting of the weak ends of these molecules. This orientation affects the surface tensions, chemical behaviour and energies of the surface molecules; and as colloids expose a maximum of such surfaces they show properties which are of a distinctive character. Thus at these surfaces loose unstable combinations with other special substances are easily formed, and colloids appear in consequence to have a peculiar and almost mysterious selective action for other substances. The phenomenon is called "adsorption"; the selected substances being adsorbed at the colloid surfaces. Colloids are thus used in many industrial processes to separate other substances from each other, to remove impurities, and in other ways to act as a selective separator of mixed substances. They also act as catalysts; that is to say, at their surfaces chemical actions take place and combinations are effected which otherwise would not be brought about. The colloid surface is apparently a special field of force or influence, in which other chemical or physical reactions besides selective adsorption are facilitated. The enzymes, for instance, in the protoplasm of the cell are complex chemical substances in very minute colloid form, with the surface molecules or radicles specially orientated so as to facilitate in a most marvellous way the chemical and physical processes which are necessary for the organic activities of life. But enzymes are very particular in their action, and each particular process has its own particular enzyme to bring it about. Thus enzymes transform the sugar or sugar-like contents of certain plants into alcohol; but each plant has its own enzyme, which will only operate on the material of that plant. Similarly chlorophyll is a complex chemical compound probably in colloidal dispersion in the protoplasm of leaf cells and other green cells, and its colloidal surfaces are "fields" in which the energy of sunlight can synthesise the carbon dioxide of the air into organic compounds which ultimately take the form of sugar, starch and cellulose. No laboratory has ever been able to make sunlight perform this wonder; but the colloidal surface "field" of chlorophyll can do it, and in that way provide for the sustenance of all organic life on this globe.

The marvellous behaviour of matter at its surfaces in the colloid state, and especially its mysterious "selective" power, has raised the hope that here the bridge may yet be found between the inorganic and the organic. Thus Dr. S. F. Armstrong says: "Enough has been said to show how the conception of an orientated active structure at the surface of a colloid aggregate might endow it with selective power of so fine a nature as almost to merit the description of intelligence; the further prosecution of research on these lines may well serve to bridge the gap between us and the full understanding of vital activity."[1]

It has been usual to distinguish "physical" from "chemical" combination. The New Physics has, however, made it clear that there are two types of chemical change,

[1] *Chemistry in the Twentieth Century*, p. 17.

involving two types of chemical combination and structure. The one type, which prevails among the salts, acids, and bases of inorganic chemistry, is a much looser, less rigid combination or union than the other, which prevails among the carbon compounds of organic chemistry. Thus common salt, which is a combination of sodium and chlorine, is now understood to be a more or less loose aggregation of free positive sodium atoms or ions held in equilibrium by an equal number of free negative chlorine atoms or ions; the equilibrium being fairly stable, without any actual union of the atoms such as was assumed by orthodox chemists. In organic compounds, however, the linkage of the constituent atoms is real, and the compound is not a system of free ions in equilibrium, but a real combination or fixed structure of the atoms concerned. Organic compounds thus display an advance in respect of chemical structure in substances. While in inorganic salts and similar substances the looser arrangement of the atoms or ions approximates to the type of "physical" combination, in organic substances, on the other hand, the chemical union is more thorough and intense, and leads to a closer structural character, linked together by common electrons. In this connection it is important to remember that organic compounds are the mechanisms of life: we may therefore say that as we approach life we witness a more intense element of structure in chemical compounds.

Life may have arisen in—at least it now uses as its mechanisms—chemical substances of a subtler structure than that which characterises inorganic compounds.

In connection with the explanation of the structure of the atom given above the question arises whether the structure of the atom is really as above indicated, or whether we have merely to do with a hypothesis to explain the known facts. The question is important, because it raises one instance of the general method of scientific explanation. Science deals with sensible phenomena and tries to co-ordinate them in accordance with known physical laws, and in doing so has often to interpret the sensible phenomena in a particular way in order to effect the necessary intelligible co-ordination. Thus, in the case of the atom, its existence as a fact is no longer disputed, but its structure on the model of a planetary system is no more than an inference from well-grounded sensible phenomena; and we cannot, therefore, say for certain that the above is the actual structure of the atom. The sensible phenomena are quite different from the inferred structure, but they are quite definite, and have been most minutely measured or calculated. The electron and the nucleus have not been observed, but certain light effects, which they accurately express, have been observed, and from these effects their mass and other properties have been calculated. The sensible phenomena actually observed include light effects, which are explained on the hypothesis of their transmission in particular wave-lengths; these explanations accord with the observed effects, and again form the basis of the supposed velocities, rotations and orbits of the electrons and nuclei, which are not directly observed but calculated with extraordinary minuteness and accuracy on the basis of the observed light effects. Similarly the light from the atom comes in definite observed quantities, which it has hitherto only been possible to interpret intelligently as sudden changes in the orbits of the rotating electrons. The observed phenomena are light effects of various definite qualities and quantities; the rest is theory or hypothesis, in which the elements of quality and quantity in the sensible phenomena are so minutely analysed and translated into elements of time and space as to result in the structure of the atom above given. And this structure is then tested by all the phenomena which call for explanation, and it is only finally accepted when it affords a complete explanation of them all. The electrons, the nucleus, the revolutions of the electrons round the nucleus, the sudden leaps of the electrons from one orbit to the other: these are not observed realities or sensible phenomena, but they all rest on a basis of sensible light effects, which have been most meticulously determined and tested by reference to other observed phenomena. They are therefore not sensible realities but scientific realities. They are not directly observed, but deduced from observations. They are the reflection, so to say, of the sensible phenomena in the human mind with its particular conceptual equipment. And if they are not the actual forms of nature, they are so close to them and measure and represent them so completely, that for us humans they are accepted as true and correct, that is, in experimental accord with the deliverances of our senses. Thus the apparently unrelated and unintelligible data of sense in a particular case are converted by the mind into the structure of the atom; and the atoms with all their inner units and arrangements become the conceptual or scientific entities which correspond to, reproduce, and represent the data of sense. In other words, the conceptual or scientific order arises on the basis of the sensible order, and as long as the two are in complete accord we accept them both together as the explanation of Nature. While thus according complete respect to both orders, we should always bear in mind that the sensible order is the governing factor to which the conceptual order has to conform. As long as it does so conform we accept it, not as sensible reality, but as an accurate measure and expression and completion of sensible reality. The hypothetical structure of the atom reproduces and expresses the observed facts; without such structure the observed facts are unintelligible and inexplicable. We therefore accept the structure as a true and accurate explanation of the observed facts, even though it has not been directly observed as a structure.

I conclude this chapter with a few general reflections on the nature of matter which will serve to emphasise and interpret the results of the foregoing discussion.

As indicated in the first chapter, the object of this work is to make a modest contribution towards the reform of the fundamental concepts of matter, life and mind, to assist in breaking down the apparently impassable gulfs between them, and to interpret them in such a way as to present them as successive more or less continuous forms and phases of one great process, or as related progressive elements in one total coherent reality. In pursuance of that general object my aim in this chapter is to pave the way for a reform of the concept of matter, to break down the old concept of matter as something inert, passive, barren, dead, as something with absolute contours and nothing beyond, as something presenting an impassable barrier to the kingdoms of life and mind beyond. This cannot be done by general philosophical reflections on the nature of matter as an object of thought, nor by launching a general invective against it, but only by a careful consideration of the concept of matter by the light of all the available physical knowledge. This must be my excuse for having referred to the Relativity Theory and the New Physics at some length. Certain general results emerge from our discussion which have an important bearing on the concept of matter.

In the first place, the old concept of matter as dead, passive, inert is clearly inconsistent with the recent developments of physical science. The old contradictory notion of dead matter as the vehicle and carrier of life must disappear in the light of our new knowledge. The difference between matter and life is no longer measured by the distance between deadness or absolute passivity on the one hand, and activity on the other—a distance so great as to constitute an impassable gulf in thought. The difference between them is merely a difference in the character of their activities. So far from matter being pure inertia or passivity, it is in reality a mass of seething, palpitating energies and activities. Its very dead-weight simply means the push of inner activities. Its inertia, which is apparently its most distinctive quality and has been consecrated by Newton in his First Law, has received its death-blow at the hands of Einstein. From the new point of view the inertia of matter is simply the result of the movement of Nature's internal energies; its apparent passivity is merely the other side of its real activity. Matter itself is nothing but concentrated structural energy, energy stereotyped into structure. As space contracts with velocity, so mass or the inertia of matter increases through that contraction, and both the mass of matter and its quality of inertia or passiveness are therefore mere variable dependent aspects of Nature's high-speed energies. From this point of view matter is but a form of energy, concentrated by its exceeding velocity, and structured to appear massive or substantial. The very nature of the physical universe is activity or Action. The Law of the Quantum rules all.

The repercussion of all this on the old concept of matter is deadly. Once the new point of view is thoroughly realised and assimilated into popular thought, the bugbear of matter will cease to trouble our peace. We shall no longer continue to stare at a hopeless irreconcilable contradiction in experience. With the dissolution of the old traditional concept of matter the dead-weight of its utter passivity will disappear from men's minds, and one of the greatest partition walls in knowledge will fall down. The contacts with life may still be very difficult to establish. But at any rate the impassable gulf will have disappeared. With the contours of matter razed, its field will itself point the way for the transition to the kingdom of life beyond. For the fields of matter and life will overlap, intermingle, and interpenetrate each other, the fruitful contacts will be established, and the enriched and broadened concepts of matter and life will appear as what they are—different phases in the evolution of an essential unity. The breakdown of the old concept of matter will have prepared the way for a great advance towards a new synthetic world-conception.

In the second place, another advance of the New Physics has perhaps even greater significance in effecting a *rapprochement* between matter and life. I refer to the effect of Radioactivity in destroying the permanence of the natural elements, and in explaining the genesis of the elements from one another. Radioactivity has done a somewhat similar work for matter as Darwin's theory of Organic Descent did for life two generations ago. The fixity of the types of matter has followed the fixity of the types of life to the limbo of the obsolete. Of course there are marked differences in the operation of Radioactivity and Organic Descent. In one respect Radioactivity has not proved as powerful a factor as Organic Descent, for it holds out no promise of the creation of new species or elements beyond those already known. The Periodic Table does indeed indicate the vacant places for half a dozen more guests yet to arrive. But the number of elements is definitely and narrowly limited, and we have no reason to look forward to any large increase beyond those already known. In another important respect Radioactivity differs from Organic Descent. Organic Descent professes to show how new and future species arise through variation and selection from those already existing. Radioactivity operates in the opposite direction and indicates how by elimination of certain unit constituents from a complex element there may be established a regress to another simpler known element. In the time-series Organic Descent professes to move forward, while the process of Radioactivity appears to be backward, or to retrace evolutionary steps taken in the past. In still another respect Radioactivity appears to be even more effective than Organic Descent, for it exhibits before our eyes the process of the transmutation of elements, while it is not yet definitely established that any natural species has yet been raised in the laboratory or will ever be raised in any period of time short of geological periods. In a final respect there is a striking similarity between the two factors in that they both appear to proceed by definite substantial increments or decrements in effecting transmutations. Radioactivity expels definite numbers of Helium nuclei as steps in the transmutation of elements. According to De Vries and others the process of advance from old to new species or varieties is by way of definite marked mutations, and not by the slow summation of minute discontinuous variations. And the present-day Geneticists emphasise this similarity still more by identifying all organic variations with differences of chromosomes or genes in the nuclei of varying or mutating species.

The above differences in the operation of the two factors of Radioactivity and Organic Descent arise partly no doubt from inherent differences between matter and life, but also partly from other possible differences in their circumstances of a less fundamental character. Thus life is a mere child on this globe and is yet in the heyday of its growth and increase. As yet it recognises no limit or barrier in its first flush of youth. It spends with a lavish prodigality, which is in striking contrast to the frugality and conservatism of matter, for which the laws of Conservation and of Least Action have become the last word of wisdom and the unbroken rule of action. But then matter is old, old as the beginning, so old that its wrinkles are the fundamental curves of the Space–Time universe. Life has only just begun, since the yesterday of Eozoic times, in the upbuilding of its new forms and types, and in this task it can proceed for millions of years to come. Matter, on the contrary, had completed its active race probably more than a thousand million years before life began. It had built up slowly and laboriously in nebular and solar heat, and amid conditions beyond the possibility of our knowledge or imagination, the elements from their simplest to their most complex forms, and from these again substances and compounds in rising complexity until at last protoplasm was reached. And in the favouring bosom of protoplasm life could be nurtured from its simple chemical beginnings and launched on its great career, most of which is still before it. The work of matter is done; in the great Space–Time curve it is now regressing from the more complex to the simpler types or elements, just as in organic Evolution we see a tendency for the most highly evolved and differentiated types to hark back for stability to simpler and stronger types. Radioactivity is doing to-day what Organic Descent (when it will indeed have become a descent) will do in the fullness of its time, when Life's spirit of adventure will have abated, and its aim will be safety and conservation rather than progress.

When all allowance has been made for the differences in character and operation of Radioactivity and Organic Descent, there still remains a striking and unmistakable similarity between them. And between the Periodic Table of Chemistry on the one hand and Systematic Botany and Zoology on the other there remains something very much like a family resemblance. The concepts of orders, genera and species could be applied to both; and in both cases there is a fluidity and plasticity of types which proves that, although they are in different kingdoms, yet they are in the same world of forms and geneses. One rises from a study of the Periodic Table and the New Physics with the feeling that matter can quite justifiably claim some distant relationship with life, and that life need not be quite ashamed of the rock whence she was hewn.

The intimate character of structure which the material universe and its field discloses justifies another general observation as bearing on the concept of matter. We have already seen that, properly understood, the ideas of activity, plasticity and development apply to matter in a sense not entirely dissimilar to that in which they apply to life. I am going to make a more daring suggestion and to indicate that in another even more important respect matter approximates to life. The structure of matter indicates that matter is also in a sense creative—creative, that is to say, not of its own stuff, but of the forms, arrangements and patterns which constitute all its value in the physical sphere. It is creative in a sense analogous to that in which we call life or mind creative of values. Remember that according to the new point of view we have not to judge of matter from the outside and as indifferent external spectators. We have to identify ourselves with the point of view of matter, so to speak. We have humbly to get into that closed cage; we have to take our post on that plane circular rotating disc. We have to interpret matter from the inside, from a point of view which is that of matter and not remote from and indifferent to it. And from that intimate angle matter is seen to create its structures and patterns and values very much as life or mind does on another much higher plane. Hitherto the idea of creativeness has been confined to the organic and mental aspects of the universe. Those who have called the universe creative have implicitly referred to the activity of life and mind in creating new arrangements, meanings and values. It has not been suggested that, from another point of view, the physical universe is also creative. The principles of the conservation of matter and energy have effectively barred any such idea. Novelty, originativeness and creativeness are quite inconsistent with the ordinary point of view and the popular ideas of matter as well as the more rigid mechanistic conceptions of science. Nobody, however, could have followed the above exposition of the structural character of matter without beginning to appreciate that in its evolution or creation of the forms, structures and types which characterise it from beginning to end, matter or the physical element in the universe is in a sense as truly creative as is organism or mind. The "values" of matter or the physical universe arise purely from these structures and forms. If the stuff of matter or energy or action were not definitely structural but diffuse throughout space, the entropy of the universe would be absolute, and its value for this cosmos from all points of view would be nil. The efficiency, utility and beauty, in short the values of matter, arise from the structures which are the outcome and the expression of its own inherent activities. In a very real sense the idea of value applies as truly and effectively in the domain of the physical as in that of the biological or the psychical. In both cases value is a quality of the forms and combinations which are brought about. Whether they are structures resulting from the activities of matter, or works of art or genius resulting from the activities of the mind, makes no real difference to the application of the ideas of creativeness and value in either case. Once we get rid of the notion of the world as consisting of dead matter, into which activity has been introduced from some external or alien source; once we come to look upon matter not only as active, but as self-active, as active with its own activities, as indeed nothing else but Action, our whole conception of the physical order is revolutionised, and the great barriers between the physical and the organic begin to shrink and to shrivel. Organism has by its inner activities and the influences of the environment evolved its own forms and types, and this great life-process is still going on before our eyes. As I have already suggested, a similar evolution of material structures and elemental types may have gone on during the practically infinite period of past time. And it may even be that, although new elements will no more be evolved, derived structures are still being created under suitable conditions. It is interesting to note, for instance, that under novel laboratory conditions, new substances are continually being synthetically produced. The whole romance of the Aniline dyes is a tribute to the still active "creativeness" of matter under the proper external conditions.

These considerations, in so far as they have any force, must influence our concept of matter and tend towards reducing the utter heterogeneity which marks our traditional concepts of matter and life. Of course a great difference remains between these two concepts, between the chemical compound on the one side and the organic cell on the other. It would be futile to attempt to argue away this difference. It is and remains great, but its character has been fundamentally transformed. We may put the conclusion of our discussion in this way. In organic Evolution we come across mutations—not absolute breaks with the past, but sudden long steps of advance on the past, where one species or variety leaps forward from and in advance of another. In the advance from matter to life there is a leap forward, not as between species, but as between kingdoms. And we may conclude by saying that, instead of the old impassable gulf between matter and life, between the chemical compound and the cell, we have found on closer scrutiny only a mutation— the greatest mutation of all undoubtedly in the whole range of science, but essentially nothing more than a mutation. They present the faint lineaments of a family resemblance, and as science advances and our philosophy looks more deeply, the resemblance will become clearer and more unmistakable.

Lastly, we have seen that matter in its colloidal state discloses properties and shows a behaviour which seem in some way to anticipate the processes and activities of life in its most primitive forms. In any case it begins to lay the basis of those physical and chemical reactions which are specially required for vital activities. It shows a certain power of selectiveness, which may be related to chemical affinity, but which seems to have a farther reach and to partake of the character of life. It begins to manufacture substances, such as chlorophyll and hæmoglobin, which are the special mechanisms of life, and without which life as we know it could not be. These substances are the links which connect material structure with the life structures which are to follow in the course of Evolution. They are themselves inorganic chemical substances, but they are the special instruments and the very basis of life, so to say. At their colloidal surfaces the energies of Nature are utilised to convert the inorganic material of Nature into the most complex organic substances required for the sustenance of life; and the conversion is brought about by processes which, however simple and direct apparently, have hitherto defied all attempts at imitation in our most highly equipped laboratories. We therefore see matter in this colloidal state reaching up to the very threshold of life, so to speak. A gap remains; a great leap may have taken place across it. But beyond a doubt some forms of matter in their colloidal state are fairly close to life in their properties. And it may even be that life began with much more primitive forms and structures than any of which we have knowledge to-day. Thus the gap may not have been so wide nor the leap so great as would appear to us to-day.

CHAPTER IV

THE CELL AND THE ORGANISM

Summary.—The cell is the second fundamental structure of the universe. It is possible that both before and after the origin of atoms and cells, as well as in between, other structures arose in the course of cosmic Evolution. If so, they have passed away, and we have now only these two permanent survivals which we can scrutinize for clues as to the basic character of the universe.

In the study of animate nature Evolution or Organic Descent has till recently attracted most attention. But more recently the study of the structure and functions of the cell has come rapidly to the front and now probably forms the principal centre of interest in Biology.

That all plants and animals consist of cells; that cells contain certain peculiar bodies called nuclei; that all higher organisms arise from cell-fusions in which the nuclei play a prominent part—all these facts have been discovered only in comparatively recent years; and our knowledge of cells is therefore still in its earliest stage. But Cytology is now, with much-improved methods and appliances, making rapid strides, and great discoveries are confidently looked forward to.

Besides the nucleus the cell consists principally of a rapidly circulating jelly-like fluid, enclosed in a more or less well-marked wall or membrane of a permeable character; and the fluid contains numerous exceedingly complex chemical compounds in solution or in the colloid state. The structure of a cell is therefore most complex, and in fact comparatively little is yet definitely known about it. Its functions are even more mysterious, for they include practically all the activities which we see in developed organisms—birth, growth, breathing, feeding, digestion, self-healing, reproduction, and death. Its most distinctive function is metabolism, which means that it thoroughly alters and transforms all food materials before assimilating them; that all its apparently physical activities are of this transformative metabolic character instead of being simple mechanical operations. It appears to form complex chemical compounds, called enzymes, which in their colloid state enable these distinctive radical transformations to be effected. The apparently simple physical processes such as osmosis in the cell are really much more complicated, as they are effected through enzyme action, which is a physico-chemical mechanism distinctive of organisms. The laboratory attempts to repeat organic processes throw, therefore, little light on the exact nature of these processes.

The origin of the cell is the origin of life and is still a profound mystery. However, the reproduction of cells seems to admit us to the inner secrets of life, and the cell-divisions which precede cell-fusions in reproduction have an extraordinary semblance to electrical situations, and seem somehow to connect the electrical structure of the atom with a possible electrical origin of the cell. It is now, however, impossible to follow up this clear semblance further, as the original electrical processes have probably become overlaid with other developments which have transformed them.

Judging from the action of sunlight in the growth of plants it is not improbable that the cell of life arose when the sun was both warmer and richer in chemically active rays, and when the waters of the earth still contained many substances in solution and colloid dispersal. The adhesion of cells to each other would account for the origin and development of multi-cellular organisms; and the divisions of cells, which we now see in growth and reproduction, may have arisen originally from the breakdown of cells or groups which had become too complex to be stable.

The reproduction of plants and animals, including the reduction division of the sexual cells, follows largely the same plan; and it is therefore probable that this wonderful organic mechanism was evolved before the bifurcation of life into the plant and animal forms took place, and thus dates back to the early beginnings of life on this globe. The plant type arose from its dependence for food on air and earth, which was consistent with fixed positions; while animals, needing organic foods, required mobility, and in consequence developed a motor system with a nervous system to work it, and ultimately a brain to direct and control it.

The cell differs from the atom or molecule in its far greater complexity of structure and function, in the differentiation and specialisation of its parts and organs, and in the system of co-operation among all its parts which makes them function for the whole. This co-operative system exists not only in the single cell but among the multitudinous cells of organisms. The system of organic regulation and co-ordination among an indefinitely large number of parts which makes all the parts function together for certain purposes is a great advance on the system of physical equilibrium in atoms and compounds, and is yet quite distinct from the control which, at a later stage of Evolution, Mind comes to exercise in animals and humans. Mind as we know it must therefore not be ascribed to the cell or the lower organisms; but organic regulation seems on that lower level to be even more effective than Mind is at a later stage.

IV THE CELL AND THE ORGANISM

This organic regulation and synthesis of functions is seen not only in all the ordinary functions of organisms, but more especially in their capacity for self-restoration in case of mutilation. In such phenomena there seems to be something more in actual operation than merely the parts; the parts appear to play a common part and to carry out some common purpose or to act for the common wellbeing. They seem to respond to some central pressure. There seems to be a central regulator. We have seen a factor in matter making for structure; we now see a factor in organism making for central regulation and co-ordination of all parts. We are evidently in the presence of some inner factor in Evolution which requires identification and description. That will be attempted in the next chapter.

THE atom and the cell are the two fundamental structures in the universe that we at present know of—the atom being the unit of the world of matter, the cell the unit of the world of life. In the last chapter we considered the structure of the atom and showed how the external properties of the atom were the expression and resultant of its internal energies and their structural grouping inside the atom. We saw the atom as a little complex world of its own, underlying the outward properties as well as the field of that little world. We now pass on to consider the vastly more complicated little world of the cell and its field. In the science of life the two most significant conceptions are Evolution and the Cell, the one being the unit structure and the other the general character and trend of the activities or functions of life. Round the investigation and development of these two governing conceptions most of the progress and interest in biological science since the middle of the nineteenth century have centred; and the results hitherto obtained have been most important, and practically revolutionary for our entire world-conception. And the end is by no means in sight yet. In the first chapter we saw that there were still deep-seated misunderstandings of the nature of Evolution, and that a proper appreciation of Evolution would mean a recasting not only of biological concepts but also, and above all, of our concept of matter. Let us now turn to the cell as the other and no doubt the real governing factor of the situation of life, and see what light it throws on the nature and concept of life.

A few introductory words in regard to the history of our knowledge of the cell may not be out of place here. It will be seen that accurate information even of what little we do know about the cell is of very recent date, and that we are only at the beginning of what may yet prove a great story.

In the second half of the seventeenth century Robert Hooke observed with the crude microscope then in use that cork and other vegetable substances had a vesicular appearance and he called the apparent cavities "cells." A few years later Grew and Malpighi independently observed in plant tissues these same cavities filled with fluid and surrounded with firm walls, as well as what appeared to them to be tubes likewise with walls and filled with fluid. Towards the end of the eighteenth century Treviranus showed that these tubes were cells placed in a row and elongated in the direction of the row and with the partitions between them lost. Then followed in 1831 Robert Brown's great discovery of the nucleus in the cell in plants, and in 1838 Schleiden's elucidation of the great part which the cell with its nucleus plays in the structure of plants, and shortly afterwards the application by Schwann of the new knowledge of the cell to the structure of animals also. Both Schleiden and Schwann attached great importance to the cell wall and looked upon the cells as having crystallised out of some mother substance. The contents of the cells Schleiden called vaguely "vegetable slime"; and it was not till about the middle of the nineteenth century that the great German biologist von Mohl correctly explained the contents of both vegetable and animal cells as nucleated masses of what he called "protoplasm," which was not a chemical crystallisation from other substances, but always came into being as the offspring or daughter cells from other pre-existing cells. Hence arose the formula: *omnis cellula e cellula*. This paved the way to the correct understanding of sexual fertilisation as the union of two cells, the discovery of cell-divisions, and the part played by the nucleus with its chromosomes in these divisions, and of the origin of embryos through repeated cell-divisions. And finally a concentrated effort was made by many investigators in many countries to discover in cell divisions and fusions, and especially in the part played by the nucleus, the physical mechanism of heredity. During this century the re-discovery of Mendelism by De Vries and others, and the rise of the new science of Genetics, have led to redoubled efforts to find the explanation of the many peculiar phenomena of heredity in an analysis of the parts played by the nucleus and the other elements in the protoplasm of the cell, and at present experimental cytology is being vigorously prosecuted with numerous improved methods and appliances.

Let us now consider the structure of the cell and the part it plays in organisms. I shall only summarise its most general and outstanding features, with a view to illustrating the considerations and speculations which will be advanced later. I am trying to find concepts for vital phenomena, which will be coherent not only with those phenomena but also with wider aspects of knowledge and reality, and a reference to the scientific facts and results is therefore necessary. The time is past when a philosophy of life could be evolved without a knowledge of or reference to the scientific facts and view-points.

All plants and animals consist of cells, these cells being again usually composed of various chemical substances, some of which have a very complex constitution. The number of cells in an organism varies according to its size and complexity, some of the lowest, most primitive organisms being unicellular or composed of comparatively few cells, while at the other end the higher plants and animals may contain untold millions of cells. The human brain alone is estimated to have about 9000 million cells! These cells again are of a most diverse character, the cells which build up the various parts and organs of the body being different from each other. Thus the cells of the nerves and the bones and the muscles and indeed of all parts of the animal organism differ markedly from each other, and the number of the different kinds of cells that go to the making up of a body may be indefinitely large. All these almost innumerable cells of all kinds and degrees of differentiation and complexity are arranged in a stable, orderly structure in the plant or animal body; and this structure is not stationary but in continual movement and development. The structural order which we have seen characterising the inorganic element or compound is even more characteristic of the vastly more complex organic body with its continuous mobility and transformations.

A plant or an animal can be considered from the point of view of its structure or its functions, that is to say, the activities performed by the structure as a whole or the parts of which it is composed. Viewing it merely as a structure we see the same orderly combination and arrangement of parts as in the inorganic body, only the constituent parts and the structural arrangements are far more complex than in the inorganic body. In water, for instance, or any other chemical compound, all molecules are more or less the same, and the body consists simply of a repetition of the fundamental molecule, and the structures in which the molecules are arranged are likewise of a repetitive character; while in an organic body there may be an indefinite number and variety of cells, and the varieties of arrangements and structures according to which these cells are combined in the several parts and organs of the body may also be indefinite in number.

But the difference between inorganic and organic bodies lies not only in their structures, but even more in their functions, especially the functions of the organic cells, to which there is apparently nothing corresponding in the inorganic world. About these cells we at present know comparatively little except that their functions and activities are the basis of the functions and activities of the organisms which they compose, all being co-ordinated into a single system of a new type called "life." In the march of Evolution from the inorganic to the organic the cell is the real innovation, to which nothing corresponding in the inorganic has yet been discovered. To use a metaphor, the cell is the point where matter or energy aroused itself from its slumbers and became active from within, with activities and functions which reveal its inner character and nature, so to say. It is a new structure in which energy becomes or is transformed into a new form of activity, becomes functional, becomes in some inexplicable way endowed with a power of self-help and self-control, with special characters of selectiveness and reproduction, which constitute a unique departure in the universe.

Let us summarise briefly some of the points that are known of the structure and the functions of the cell; and as the plant cell is simpler than the animal cell let us take that as the type. It consists of chemically very complex substances called in the aggregate protoplasm, which is the physico-chemical basis of all forms of life. Comparatively little is known of its composition or chemical structure. In the plant cell (less so in the animal cell) it secretes a containing wall or membrane for itself from which the cell derives its name. Inside the wall the protoplasm appears as a jelly-like fluid and consists principally of a small nucleus, which contains certain chromatin bodies of a rich protein character, and of a larger body of cytoplasm surrounding the nucleus and reticular in structure, that is to say, consisting of a network of spaces which contain various cell-saps and solutions and even minute particles of crystals and other inorganic bodies. The whole constitutes a colloidal system, as we saw in the last chapter. The cell walls are semi-permeable, admitting of the osmosis through them of certain substances and not of others, so that suitable food and other substances can be passed through the cell walls from one cell to another. There is a constant circulation and agitation of the cell fluid, which gives it the appearance of a stream, and is much more than the usual promiscuous Brownian movement in inorganic colloidal mixtures. The movement of protoplasm, whether it is Brownian or something different, has much more of the character of definite specific direction; and this is probably only an expression of that selectiveness and directiveness which are inherent and universal characteristics of all life-forms. Although little is definitely known of the details of cell-structure, the functions it performs are so many-sided, delicate, and complex that one may safely say that the cell must have an immensely complex organisation, and that the details of its constitution may never be fully known or even adequately pictured by the human mind. It represents, at the one end of the scale of existence, a minute detailed complexity which is in some sense comparable to the wonders of the astronomical universe at the other end. And all this intricate and complex little system is maintained in a state of active, moving equilibrium; it is dynamic through and through and incessantly active in all its details, and its almost innumerable activities are finely adjusted to each other and co-ordinated into a harmonious process, which not only maintains its balanced functioning for its individual life, but increases and improves it in the duration of innumerable generations. Looking at this baffling mystery of active, continually changing and developing organisation, with its continuous delicate readjustments of innumerable moving parts into one co-ordinated forward movement, we find that ordinary physical categories of description fail us. We feel in the presence of an entirely new phenomenon, which we call life, and we may even feel tempted to go further and to say that the cell has not only life but also mind. To do so would, however, be going too far, as I shall explain later.

To appreciate the position more fully let us look at some of the functions of the cell. It is very difficult to realise it, and yet it is the fact, that the little microscopic or ultra-microscopic cell probably does all or most that the plant or animal is known to do. It literally breathes or respires; it takes in, manipulates, digests and assimilates its food; it reproduces its kind; it grows, decays and dies; it heals itself when sick and restores itself when a breakage takes place. It develops special means and mechanisms to assist it in carrying out these operations, and it co-ordinates and regulates all its manifold activities in a way which implies some wonderful central control of all these functions. Let us look at these operations with a little more detail.

Unlike any other substance in nature, the protoplasm of the cell is vitally active and is in an incessant process of real creative change; parts are continually being destroyed and replaced by new protoplasm which is continually being formed. No other substance has this power of making its own material, so to say. A crystal, for instance, builds itself up from its own material already existing in solution without any change being made in its constitution. The crystal serves merely as an attractive centre round which its material, already present in dissolved form, may be deposited in solid form. With protoplasm the process of growth or renewal is quite different. The material taken in is entirely altered and recombined into the substance of which the protoplasm is composed, and this material, so altered and transformed, is then by some yet unknown process taken up and assimilated into the protoplasm or living substance of the cell. This complete transformation and this mysterious assimilation of its material is one of the most unique functions of the cell, and its far-reaching significance will later on be more particularly stressed.

The technical name for this complete transformation which the cell effects in the material it takes in is metabolism, and it may therefore be said that metabolism is the process which above everything distinguishes living from non-living matter. The cell is not a static or stationary organism; it is for ever being built up by new material which it transforms into its substances, and it is for ever being broken down through the new cell substances which it forms and gives off in order to build up the various parts of the complete plant. And the activity by which the material is taken in in one form, then transformed and assimilated into the substance of the cell, and then again given off as different cell substances for the building up of the various parts of the plant—this activity, while apparently a series of chemical and physical processes, implies a co-ordinated system which is unlike anything seen in the purely physical or chemical domain. The physical and chemical changes seem to be merely the mechanisms or instrumentalities used by a deeper organic process, which means and does much more than the physical or chemical details which we can identify. Not only is there control and organisation of these details, but the physico-chemical agencies themselves are of a new type. For instance, oxygen and carbon dioxide appear to be taken in through the stomata of the leaves, but this is not merely a case of ordinary osmosis. Again, liquid materials in the form of dissolved salts or other inorganic substances are taken in through the roots, but this also is not a case of ordinary physical osmosis. Again, these liquids rise in the plant cells as if it were merely a case of surface tension or capillary action. But as a matter of fact these are all cases of metabolism in which subtle changes take place in the protoplasm, changes whose details are no doubt apparently all of a physico-chemical character, but whose distinctive character lies not only in these details but in the new system of control in which they are organised and regulated. The theory which has been developed to account for the physico-chemical reactions which take place in all organic change and functioning is based on the assumption of very complex substances of the nature of ferments or enzymes being formed and acting in the protoplasm. It is, for instance, through the agency of the enzymes in the protoplasm that all the secretions are formed which build up the different parts of the plant. Thus also the transformation of the carbon dioxide in the green cells of the leaves into starch is not a chemical change of the ordinary type, but is effected in the presence of colloidal catalysts like chlorophyll and other enzymes, at whose surfaces sunlight can transform the carbon dioxide so as to form successively formaldehyde, dextrose, maltose, and finally soluble and insoluble starch. Mere physical and chemical reactions have been identified. But it is quite possible that there is much more, and that the organic process behind them is much more complicated and characteristic. Again, both respiration and metabolism are processes effected through enzyme action at colloid surfaces instead of being of the ordinary mechanical character. The enzymes are thus conceived as being catalytic agents existing in colloidal form in the cells and as having at their surfaces or in their "fields" the power of transforming other substances in the presence of the energy of sunlight or electricity. They do this according to the well-known chemical and physical laws, without themselves being thereby used up or transformed. The cell has the power to build up or secrete these complex enzyme compounds; with the help of these, again, it manufactures other complex substances necessary for the plant or animal life. It carries on many other functions beyond these manufacturing processes. Throughout it seems to follow simple physical and chemical rules but on a new plan. All this will serve to emphasise how vastly complicated cell structures and activities must be. A large number of the most complicated processes are carried on, scarcely one of which the best-equipped laboratory in the world can perform, and all are carried on by a little cell which is microscopic or smaller in size!

During recent years resolute attempts have been made to repeat under artificial laboratory conditions what takes place in the living plants, and certain very interesting results have been obtained. Thus an attempt has been made by Professor Baly and others to imitate photo-synthesis in the laboratory. Light of a short wave-length from a mercury vapour lamp was made to act on water and carbon dioxide, and as a result formaldehyde was obtained and, as in the green leaf, oxygen was set free: $CO_2 + H_2O = CH_2O + O_2$. Light with a somewhat longer wave-length was made to turn this formaldehyde into simple sugars.[1] However interesting and valuable these and similar results are, it is probable if not certain that they have only a distant resemblance to what takes place in the organic process, where the physical factors of sunlight and electrical change acting in the field of colloid chlorophyll are quite different, and the chemical results are brought about by mainly different processes. From a scientific point of view, however, the laboratory work

[1] It is claimed that sunlight has quite recently been successfully substituted for artificial light in these experiments.

is of undoubted interest and importance, for the further it is prosecuted and the greater the success in the synthetic formation of organic substances, the easier it will become to differentiate clearly and unmistakably between the organic and the mechanical laboratory processes.

The origin of the cell is the origin of life, and we know nothing definite about it. But the question arises whether sufficient is not known about the cell and organic development to justify us in trying to form some general idea as to its possible origin. And here we find one set of phenomena which throw a special light on the nature and development of organisms and perhaps also on their origins. The phenomena of reproduction seem to hold the very secret of life, and moreover, bring us close to the secret of matter. And this secret common to both, jealously guarded and preserved throughout the whole range of terrestrial evolution, shows a continuity unique in science, which brings together some of the apparently most diverse facts which confront us in the world of life. How well the secret has been guarded and kept and shielded from all outside influences is evidenced by the extraordinary fact that though plant and animal life must have diverged near the beginning of things, and must through many millions of years have been moving further apart in the history of this globe, yet the methods of reproduction in plants and animals are still very much alike. The romance of the reproduction of a flowering plant, which is one of the most wonderful in the world, is practically the same, down to many details, as that of the reproduction of one of the higher animals. It is very much the same in the simplest, most primitive alga as in the other members of the rising plant series—the ferns, the cycads, the conifers and the flowering plants. When we go a step further back into the past and come to the case of unicellular organisms, which reproduce themselves not by cell-fusions but by cell-division, we come to the situation, or something very close to the situation, which must have arisen when matter first organised itself into life. For what do we see? The cell when it proceeds to divide into two, assumes the

IV THE CELL AND THE ORGANISM

appearance of an electrical and polar system; its nuclear material arranges itself in parallel bodies or chromosomes like an electrical arrangement; its centrosome (if any) splits up, as if under some unbearable electrical strain, just as is the case in Radioactivity, and two polar bodies are formed from it at opposite sides of the nucleus, from which lines of force proceed throughout the now disintegrating nucleus and cell; the nuclear bodies of the breaking-up cell divide themselves equally between the two polar bodies, and aggregate and concentrate towards them until finally the separation between the two systems is complete and the material of the nucleus and cell has split into two. The division of the cell into two cells is complete. It is apparent from this summary statement how the cell in division approximates to the character of the atom of matter described in the last chapter. Were they not in the beginning both electrical systems with their nuclei, their fields and their cataclysmic behaviour? In the cell the original electrical character of the division has become overlaid with and obscured by other factors so that the electrical character is no longer recognisable, except in the general appearance and scheme of division. But originally it possibly was electrical, as it still is in appearance. Arguing back from the analogy of cell-division to the probable original rise of the cell from inorganic matter, we may imagine the building up of very complex organic or hydrocarbon compounds under favourable external conditions, in which the influence of sunlight and other forms of electrical energy played an important part, just as sunlight in the presence of chlorophyll still plays a foremost part in the production of new cells and organic substances in plant-life. We know that millions of years ago, when life arose, the sun was much hotter than it is to-day, and sunlight contained much more of the chemically active rays which facilitate organic changes. The peculiar electrical energy of the sun may therefore have played a decisive part in the origin of life. In other words, the part which electrical changes appear to play in the process of cell-division may be somewhat analogous to the part they probably played in the original rise of the cell. The basic electrical structure of matter would thus be paralleled by the more complex electrical origin of the cell. Reproduction of the most primitive forms takes place in a fluid medium, and all protoplasm still has a fluid jelly-like consistency. It is therefore probable that the most primitive forms of protoplasm might have arisen under favourable conditions of sunlight and warmth when the warm water still contained much of the crust in solution or dispersed in small particles in colloidal form, and thus presented conditions favourable for the selective formation of complex substances, such as the predecessors of the present forms of protoplasm. Recent advances in bio-chemistry have led to the isolation and discovery in the animal body of very complex substances (such as glutathione) whose external surfaces absorb free Oxygen and whose interior undergoes the opposite change of setting free this Oxygen and absorbing Hydrogen for the building up of higher structures. Thus a continual chemical process is set going which is at the same time an electrical current from without inwards, transmitting the electrical energy of sunlight contained in the Oxygen of the air to the interior cell substances, where the Oxygen is deprived of its energy and set free, and where with this energy complex compounds are built up which again store the energy required for the nutrition and other functions of the living organism. The chemical electrical system which forms the fundamental mechanism of life in the last resort simply uses the energy of sunlight, stored in the free Oxygen of the air, for building from it the body of life; and there is thus the closest connection between sunlight and life-structures. In that connection, without a doubt, the origin of life must be sought. The cell structure having once been evolved from pre-cell structure, probably by way of prolonged trial and error or " natural selection " extending over long periods of time, its multiplication or reproduction would take place through the part that electrical tension would play. Thus the complex unstable electrical structure or organic substance, whose internal equilibrium would pass through various crises and changes in its "development," would finally tend to break up and under certain conditions proceed to divide. This original haphazard division would gradually become stabilised and standardised, so to speak, until cell-division becomes the regular basis not only of all growth but also of all reproductive processes in both plant and animal.

At first there could have been no essential difference between growth and reproduction of cells. By division one cell was formed from another, and might either remain in association with the old cell, as is the case in all multicellular organisms (growth), or it might separate from the old cell and develop on its own as an independent organism (reproduction). This simple division still remains the process of growth in all organisms without distinction, and it remains the process of reproduction in all unicellular and the lowest forms of multicellular organisms. In less primitive, more developed multicellular organisms the process of reproduction has, however, become more complex, and has altered to the union or fusion of two specialised cells to form a new cell; and in such cases another scheme, involving a double set of divisions, has taken the place of the simple division. While one of these simply halves the total contents of the old cell as between the two new cells, the other or reduction division separates out the individual elements in the contents so that each of the two new cells has half of these elements (*e.g.* the chromosomes of the nucleus). This halving is necessary to prevent the continual and cumulative doubling of cell elements in the repeated reproduction of the same type of organism, and to keep the contents of cells of similar organisms constant. The two cells, thus reduced in all respects to half the original cells, then unite to form the new cell, which has once more the full complement of cell elements. This reduction division in reproduction is common to both plants and animals above the most undeveloped types, and we therefore seem to have some justification for the most remarkable conclusion that this phase of reproduction must have developed before the separation of plant and animal forms took place. It forms also the basis for that alternation of generations which is one of the most remarkable of all the phenomena of life. Thus all organisms which are reproduced through cell-fusion have a generation in which the cells have the single contents (after the reduction division) and another in which the cells have the double contents (before the reduction division). In the higher forms of plants and animals the generation of the single-content cell, or the gametophyte generation as it is called, is reduced to a very subordinate rôle and a short life, as it covers the short period of the gametes or conjugating cells (the ripe sperm-cells and ova) in flowering plants and in the more developed animals. The generation of the double-contents cell, or the sporophyte generation as it is called, has become dominant and appears as the developed plant or animal which we see in nature to-day. But in some divisions of plants the gametophyte generation is still of some prominence. Thus the moss plant is the gametophyte generation, the sporophyte generation appearing as a subordinate parasitic form. And in the ferns, where the sporophyte is the dominant form, the gametophyte appears as a distinct plant which is in some cases a perennial. And its relation to the fern was quite unknown until about the middle of the nineteenth century.

This generation of the single-content cell, or the gametophyte in the developed plants and animals of to-day, is interesting because it probably illustrates the well-known principle that ontogeny repeats phylogeny, that is to say, that the history of the individual organism recapitulates in its earlier or embryonic stages of development the various phases of development through which its types of ancestors have evolved in the past. Thus the gametophyte generation is a reminder of that earlier, simpler, more primitive phase in plant evolution when the more complex sporophyte generation, which is dominant to-day, had not yet been evolved through cell-fusion in reproduction. And it is even possible that the form of the present gametophyte may throw light on the particular descent of plant forms. Thus the gametophyte of the fern is a flat thallus-like plant which both in form and character reminds one of an alga. And it is quite possible that this form of the fern may give the clue of its origin from some alga-like progenitor in the far-distant past. May we not say that the prothallus of the fern appears to connect the alga and the fern, and thus to bridge widely separated epochs of the past in the evolution of plant forms? I suggest the idea merely for further investigation.

From speculations as to the origin of the cell we pass on to consider the differentiations which have taken place among cells generally, and the particular differentiation of cells which has led to the divergence between plant and animal forms. It is commonly thought that the animal forms are a later development and advance on the earlier plant forms. This idea is largely due to the fact that in the animal there has been the special development of the new factor of mind which, rapidly rising through the higher animals, has reached its highest level in the human individual. But although animal forms may have developed farther and come to attain to much higher levels than the plant forms, the question of origins stands on a different footing. And the evidence points rather to a common origin and to the earliest cells of life having been common to both plants and animals. Thus the lowest forms of cell life are even now practically indistinguishable into plant and animal. And it is probable that this common phase, prior to differentiation into plant and animal forms, must have lasted a very long time and have been marked by considerable advances in the development of the common cells, especially in view of the probable fact, already noted, that sexual reproduction of a fairly advanced type had probably been reached before the bifurcation took place and plants and animals were launched on their separate careers. What advance in cell development had been reached before this bifurcation it is impossible to say, as only the very lowest unicellular organisms of a common character still survive, and the geological record has no evidence to give. Differentiation in cells must have commenced as soon as the daughter cells began to adhere to the parent cell and multicellular organisms were formed. In the unicellular Pleurococcus, which is about the simplest plant form known, noticed as the green slime on the damp bark of trees or wooden posts, we see the beginnings of this process of cell aggregation, as daughter cells adhere to the parent cell until several divisions have taken place and only then separate into individual cells. The Pleurococcus cell is globular, but during this attachment the cells are flattened at the surfaces of contact. In the multicellular organisms a layer of cuticle covers the outer cell walls in contact with the air and retards the loss of water. Step by step other differentiations appear, and the plant body becomes more complex as we advance from alga to fern, and from fern to the higher seed-bearing plants. The differentiations into various organs, such as the root, stem, leaf and reproductive organs, are simply means towards the division of physiological labour. Thus in cells away from the light photo-synthesis is impossible, and they become dependent on the outer green cells; similarly the roots underground become dependent for starch on the green cells and in return absorb dissolved salts for supply to the rest of the plant. The water requirements render necessary the fibrovascular cell system, while the reproductive functions become confined to special organs. All this differentiation means more organisation and a more elaborate structure of the plant. In addition to this division of labour the struggle for existence tests the structure in other directions, and means more modification in response to the stress of the struggle and the stimulus of the environment generally. As a result the plant structure comes to be elaborated and adapted to the inner and external demands upon it, and to assume the forms which are known to us.

Besides this general differentiation and organisation, there are special causes which have brought about the divergence of plant and animal forms. A consideration of these matters falls outside the scope of our task. Generally it may be said that plant-life has been determined and stereotyped through the two processes of photo-synthesis and osmosis, the second of which has enabled it to get water and mineral salts direct from the soil, and the first of which has enabled it by the help of chlorophyll to utilise the energy of sunlight for making sugars, starch and cellulose from the carbon dioxide of the air. The plant, being thus dependent for its food solely on the soil and the air, could afford to remain stationary and mostly fixed in the soil; while animal forms, which are dependent on organic foods, have had to be mobile in order to look for and find the necessary plant or animal substances on which to live. The struggle for food has been a much harder one for animals, which have in consequence not only to be mobile and develop a complex motor system, but also to evolve in the nervous system a special co-ordinating mechanism with which to work the motor system. This mechanism, again, has led to unique developments in the direction of sensitiveness and consciousness, which in the case of man have come to overshadow all that has gone before. But mind is a later development, the discussion of which should not be raised in connection with the cell. The primitive cell of life is on the way to Mind, but Mind at this stage is still far off, and those who ascribe Mind or even potential Mind to the cell open the door to the most serious confusions. The cell undoubtedly presents a great mystery. And there is a strong temptation to ascribe its surprising activities to an inner mentality or organic psychism. But even the most highly evolved human intelligence finds it difficult to understand all that goes on in the cell. If psychism is the key, we should have to ascribe to the cell so large a measure of mentality as to reduce the whole supposition of psychism to absurdity. The cell has not yet mind. Mind as we know it, or anything at all resembling it, is a much later development in the process of organic Evolution, as will be shown in Chapter IX.

Enough has been said about the structure and the functions of the cell to give a rough general idea of what the cell is. Let us now pass on to consider the inter-relations of elements in the cell, and among cells in the same organism, and especially the aspect of co-ordination in and among cells.

In the first place I ask: Is the cell and are cells in an organism a co-operative system, in which the parts and their functions are so ordered and arranged that they co-operate for common purposes, and do not merely subserve the separate ends of the individual parts?

There could be no doubt as to the answer. The whole meaning and significance of Metabolism is that the activities of the cell are not self-centred or self-regarding. The cell functions for other cells and for the plant as a whole. One element in the cell functions for other elements and for the whole cell organism. The secretions formed in one cell are intended to build up other cells or to serve the plant as a whole. The fibro-vascular cells carry liquid food from one part of the plant to the other parts. The carbohydrates formed in the green cells are transmitted and stored as food for all the other cells; the woody substances secreted from them are meant to strengthen other cells and the plant as a whole against the forces of the environment; the aroma and bloom which are secreted from them are meant to render attractive and adorn other parts of the plant with a view to the preservation of the plant species as a whole. Indeed, all the processes of Metabolism go to prove that the plant is one vast co-operative system, in which the individual cells in their continuous functions and labours make their contribution to the common effort, and work so that other cells or the plant itself or the species to which it belongs may live. The cell in its normal structure and functions is the very type of co-operative action.

So far I believe we are still on firm ground in our description of cell activities, and the co-operative character of organic functioning will be generally admitted. Can we go further and characterise this co-operation more closely? What is the nature of this cell or organic co-operation? Is it spontaneous or controlled? And if controlled, is it controlled internally or externally? Let me repeat the question in another form. Are the cells and the organs which they form in the same plant or animal free and independent, so that the co-operation which we observe in their functioning is a mere accidental result of their individual uncontrolled reactions and behaviour? Or is there some co-ordinating factor which influences the cells and their organs in some specific direction, and thus co-ordinates and unifies their functions and produces the co-operation we observe? And if the cells are not independent agents in the make-up of the organism but are under some form of unifying influence or control, is their apparent co-operation due to an external factor, like Natural Selection as commonly understood? Or is there some internal element of co-ordination, the influence of which is felt by the different cells, and in response to which they react, so that their functions proceed generally on the lines of a plan or pattern given by the nature of the particular organism? In either case there would be co-operation on the basis of co-ordination, but in the one case there would be an external, and in the other an internal, factor at work in this co-ordination. It will be seen that the issue here raised as between the cells inside the organism is analogous to that which Darwinism has raised as between separate organisms in their struggle for existence. The answer, so far as the struggle among organisms is concerned, will be discussed in Chapter VIII. And the results there reached will probably apply also to the case of the cell or the cells in an organism which is here raised. The subject is not free from controversy, and in this chapter I wish to avoid controversy and simply to describe the facts in the ordinary language of metaphor which I trust will not prove misleading. And looking at the facts in an unprejudiced way, and without a bias in favour of any particular theory, one cannot help being struck by the way in which the cells in an organism not only co-operate, but co-operate in a specific direction towards the fulfilment and maintenance of the type of the particular organism which they constitute. At this stage we have to steer clear of all ideas of plan, purpose or teleology in the organic procedure. But, even so, the impression is irresistible that cell activities are co-operative, that they are inherently or through selective development co-ordinated in a specific direction, and that the impress of the whole which forms the organism is clearly stamped on all the details. The case is utterly unlike that of physical forces, which are alike, which are repetitions of each other, and which can be added or subtracted or otherwise expressed arithmetically. The cells are different, they are differentiated in definite directions, and the totality of differentiations fits into a plan or scheme, the fulfilment of which constitutes the complete organism. There are no repetitions, there is uniqueness everywhere, and the various unique entities and their functions fit into each other more or less so as to produce an organic whole, unlike any other organism. And in some indefinable way this whole is not an artificial result of its part; it is itself an action factor like its parts, and it appears to be in definite relation with them, influenced by them and again influencing them, and through this continuous interaction of parts and whole maintaining the moving equilibrium which is the organism.

Look, for instance, at the way in which organisms behave when some cells or organs, necessary for their maintenance, are removed or injured. It is well known that many plants and animals have the power of restitution in case of damage or mutilation. The newt forms a new leg in the place of the severed limb. The plant supplies the place of the severed branch with another. The regeneration may be effected from different organs and by different organs. Thus if the crystalline lens is removed from the eye of a Triton, the iris will regenerate a new lens, although the lens and the iris in this case have been evolved from quite different parts. Numerous similar curious facts of restoration could be mentioned. The broken whole in organic nature restores itself or is restored by the undamaged parts. The cells of the remaining parts set themselves the novel task of restoring the missing parts. The power to do this varies with various plants or animals, and varies also with the different parts in the same plant or animal. Generally one may say that the more highly differentiated and specialised an organism or a cell is, the smaller is its plasticity, or the power of the remaining cells to restore the whole in case of injury or mutilation. But the fact that the power exists in numerous cases is a proof that not only can the cells through reproduction build up the original organism according to its specific type, but also that when this type is damaged, the remaining cells or some of them can restore it, and recomplete the whole. The normal power of the cells to build up an organism in reproduction according to type is one thing, and it is marvellous enough even though one looks upon it as merely a case of inherited routine. But the abnormal power to do this in the very unusual case, so far removed from all idea of routine, where the type is broken down is something different, and shows how effective the power of the organism as a whole is, and how strong the tendency towards the whole is even in the individual cells. The damage creates a need, and the need stimulates the remaining parts to perform the functions of the damaged parts or to restore them in whole or in part. The very nature of the cells is to function as parts of a whole, and when the whole is broken down an unusual extra task automatically arises for them to restore the breach, and their dormant powers are aroused to action. And this happens, so far as we can see, simply as a matter of interior economy and domestic regulation in the organism itself and without previous education for the new rôle. The interaction between the organism and its cells is indeed most subtle and intimate, both seem to be active factors in the maintenance of the whole and in the restoration of any parts that may be missing and necessary for the whole. So intimate is their interaction that it is almost impossible to say where the influence of the one ends and the other begins.

The aspect of co-ordination or subordination of parts to the whole is also most significantly illustrated by the phenomena of reproduction which I have already referred to in another connection. Reproduction not only carries us back to the past and its riddles, but also forward to the future, and it is the reproductive system of organisms that we must scan most closely if we wish to understand this aspect of organic activities. For in reproduction the cell or the organism clearly appears to look beyond itself, its functions become transcendent, as far as it is itself concerned; its efforts and energies are bent on objects and purposes beyond itself. In fact, in reproduction the cell or the organism bears clear testimony to the fact that it is not itself alone, and that it is part of a larger whole of life towards the fulfilment of which its most fundamental functions are directed. As an illustration of the co-ordinated inter-relations of parts and whole in organism, nothing can therefore be more significant and important than the facts of organic reproduction. Here more than anywhere else the importance of the whole as an operative factor appears, not merely the immediate whole or individual organism, but also the transcendent whole or the type which has to be reproduced and maintained at all costs. Throughout the entire range of organic nature one is impressed with the essential selflessness, the disregard of self, and the transcendence of self in the reproductive process, which harnesses the individual to the purposes of the race, exhausts its reserves of strength, and often costs it its life. On that process is stamped, as on the very heart of nature, the principle of sacrifice, of the subordination of part to the whole, of the individual to the race or type.

The preceding analysis will have enabled us to realise that the plant or the animal is a whole consisting of millions of parts in the form of cells of all kinds, while the cells again are smaller wholes of indefinite complexity and marvellous activities. All these parts are co-ordinated and arranged down to the most minute details, and function with the most complete collaboration in support of each other and the whole organism. The organism is indeed a little living world in which law and order reign, and in which every part collaborates with every other part, and subserves the common purposes of the whole, as a rule with the most perfect regularity. It is this perfect community of functions and unity of action in a system consisting of innumerable parts and the most complex structural arrangements that makes the organism such a striking type of a whole. We have seen structural order as the characteristic of inorganic matter; we now see active co-operation and unity of action superadded as the characteristic of the organism. We admire the order and co-operation of a beehive or a community of ants; in the organism we see a more perfect order and a more wonderful co-operation in a situation which is perhaps not much less complex than either. And just as the individual bee or ant lives its own life and is not lost in the joint venture of the hive or nest, so the individual cell lives its own life and specialises and perfects itself in the organism which it helps to form and to serve without loss or sacrifice of its own identity. The organism embraces innumerable smaller organic units whose identity is not swallowed up in it, is expressed and not suppressed by it. The large organism does not only mean the union and co-operative harmony of its smaller units, but also as a rule the more perfect individuation and specialised development of these units in the harmony of the whole. The plant or animal body is a social community, but a community which allows a substantial development to its individual members. And its nature and structure are such that it can only perfect itself through the differentiation and development of the members which compose it. But while this is so, while (as we shall see more clearly in the sequel) individuation is fundamental in nature, we have to recognise that co-operation plays a no less important and fundamental part. An organism is fundamentally a society in which innumerable members co-operate in mutual help in a spirit of the most effective disinterested service and loyalty to each other. Co-operation and mutual help are written large on the face of Nature. Nay, more, if cell structure and function can teach us anything; they are imprinted deep on the nature of the universe, they are the very meaning and soul of Nature. We may travel far through the realms of Evolution, but nowhere shall we find a more perfect co-operation or a more beautiful illustration of mutual help of one part for another, and of all parts for the whole, as well as of the whole for all its parts, than in the little insignificant cell, which seems to hold the very secret of the universe. Anticipating the language of later developments, we may say that in the cell there is implicit an ideal of harmonious co-operation, of unselfish mutual service, of loyalty and duty of each to all such as in our later more highly evolved human associations we can only aspire to and strive for. When there was achieved the marvellous and mysterious stable constellation of electrical units in the atom, a miracle was wrought which saved the world of matter from utter chaos and chance. But a far greater miracle was wrought when from the atomic and the molecular order there was evolved a still deeper and subtler order in the inner co-operative creative harmony of the cell. These two fundamental structures are the great abiding achievements in the course of Evolution, before the advent of Mind, and though many other experiments were made before and in between these successes, they proved unstable and were discarded and abandoned, and are now searched for in vain. We have to scrutinise these abiding peaks of achievement if we wish to understand the real nature of the Evolutionary process, and if we wish to form an idea of the nature of the ground in between these permanent structures which has been washed away in the endless lapse of time. And when we find the two to be not utterly different but expressions of a somewhat similar inner progressive tendency of nature, and when we find later, on the mental and spiritual levels of development, still clearer expressions of a similar tendency, we shall be justified in concluding that we are face to face with something real and causal in the form of a natural operative factor of a fundamental and universal character. The impression becomes so strong that it is not so much a matter of speculation as a recognition of clear simple facts before us. The permanent structures in Nature have been and are still being patiently investigated for us by Science. As I said in the last chapter, they present more than a faint family resemblance and enable us to recognise the unity which underlies them all and to draw certain conclusions as to the origin of this unity. They point strongly in the direction of some inner natural factor in Evolution of which they are the expression. The evidence in favour of such a natural factor of a creative synthetic ordering character has been accumulating in this and the two preceding chapters; it requires isolation, identification and exact formulation, if that were possible. And in the next chapter a preliminary attempt at such an identification and formulation will be made.

CHAPTER V

GENERAL CONCEPT OF HOLISM

Summary.—The close approach to each other of the concepts of matter, life and mind, and their partial overflow of each other's domain, raises the further question whether back of them there is not a fundamental principle of which they are the progressive outcome. That is the central problem of this work.

Two conceptions of genesis or development have prevailed. The one regards all reality as given in form and substance at the beginning, either actually or implicitly, and the subsequent history as merely the unfolding, explication, *evolutio*, of this implicit content. This view puts creation in the past and makes it predetermine the whole future; all fresh initiative, novelty or creativeness is consequently banned from a universe so created or evolved. The other view posits a minimum of the given at the beginning, and makes the process of Evolution creative of reality. Evolution on this view is really creative and not merely explicative of what was given before; it involves the creative rise not only of new forms or groupings, but even of new materials in the process of Evolution. This is the view of Evolution to-day commonly held, and it marks a revolution in thought. It releases the present and the future from the bondage of the past, and makes freedom an inherent character of the universe.

Creative Evolution involves both general principles or tendencies and concrete forms or structures; philosophy studies the former, while science has more exclusively concentrated on the latter. Yet both are necessary to reality; and any universal formula of Evolution must include both the general activity or tendency and the concrete particular structure, as one cannot be deduced from the other. Bergson made an attempt to deduce Evolution with all its multitudinous forms from homogeneous, pure, undifferentiated Duration. This was, however, not possible, and he had to call in the practical spatialising Intellect to infect Duration in order to make her productive; he thus made the Intellect play a one-sided and, at the same time, excessive rôle in the shaping of the forms of the universe. It would be a better procedure to take some natural unit or sample section of Nature for our starting-point, and thus to keep as close to her and her concreteness as possible. The last two chapters give us a clue where to look for a beginning. Both matter and life consist of unit structures whose ordered grouping produces natural wholes which we call bodies or organisms. This character of "wholeness" meets us everywhere and points to something fundamental in the universe. Holism (from ὅλος = whole) is the term here coined for this fundamental factor operative towards the creation of wholes in the universe. Its character is both general and specific or concrete, and it satisfies our double requirement for a natural evolutionary starting-point.

Wholes are not mere artificial constructions of thought; they point to something real in the universe, and Holism is a real operative factor, a *vera causa*. There is behind Evolution no mere vague creative impulse or *Élan vital*, but something quite definite and specific in its operation, and thus productive of the real concrete character of cosmic Evolution.

The idea of wholes and wholeness should therefore not be confined to the biological domain; it covers both inorganic substances and the highest manifestations of the human spirit. Taking a plant or an animal as a type of a whole, we notice the fundamental holistic characters as a unity of parts which is so close and intense as to be more than the sum of its parts; which not only gives a particular conformation or structure to the parts but so relates and determines them in their synthesis that their functions are altered; the synthesis affects and determines the parts, so that they function towards the "whole"; and the whole and the parts therefore reciprocally influence and determine each other, and appear more or less to merge their individual characters: the whole is in the parts and the parts are in the whole, and this synthesis of whole and parts is reflected in the holistic character of the functions of the parts as well as of the whole.

There is a progressive grading of this holistic synthesis in Nature, so that we pass from (a) mere physical mixtures, where the structure is almost negligible, and the parts largely preserve their separate characters and activities or functions, to (b) chemical compounds, where the structure is more synthetic and the activities and functions of the parts are strongly influenced by the new structure and can only with difficulty be traced to the individual parts; and, again, to (c) organisms, where a still more intense synthesis of elements has been effected, which impresses the parts or organs far more intimately with a unified character, and a system of central control, regulation and co-ordination of all the parts and organs arises; and from organism, again on to (d) Minds or psychical organs, where the Central Control acquires consciousness and a freedom and creative power of the most far-reaching character; and finally to (e) Personality, which is the highest, most evolved whole among the structures of the universe, and becomes a new orientative, originative centre of reality. All through this progressive series the character of wholeness deepens; Holism is not only creative but self-creative, and its final structures are far more holistic than its initial structures. Natural wholes are always composed of parts; in fact the whole is not something additional to the parts, but is just the parts in their synthesis, which may be physico-chemical or organic or psychical or personal. As Holism is a process of creative synthesis, the resulting wholes are not static but dynamic, evolutionary, creative. Hence Evolution has an ever-deepening inward spiritual holistic character; and the wholes of Evolution and the evolutionary process itself can only be understood in reference to this fundamental character of wholeness. This is a universe of whole-making. The explanation of Nature can therefore not be purely mechanical; and the mechanistic concept of Nature has its place and justification only in the wider setting of Holism. In its organic application, in particular, the "whole" will be found a much more useful term in science than "life," and will render the prevailing mechanistic interpretation largely unnecessary.

A natural whole has its "field," and the concept of fields will be found most important in this connection also. Just as a "thing" is really a synthesised "event" in the system of Relativity, so an organism is really a unified, synthesised section of history, which includes not only its present but much of its past and even its future. An organism can only be explained by reference to its past and its future as well as its present; the central structure is not sufficient and literally has not enough in it to go round in the way of explanation; the conception of the field therefore becomes necessary and will be found fruitful in biology and psychology no less than in physics.

IN this chapter we approach the central problem of our inquiry. In the preceding chapters we have seen the concept of matter coming closer to the concept of life; we have seen the concept of life, in the cell, and in organism, and in Evolution generally, tending towards the concept of mind. We have seen these three fundamental concepts, at first apparently so utterly unlike and so far apart, approaching each other and overflowing each other in the real structures and evolution of the universe. The question now arises whether there is not something still more fundamental in the universe, something of which they are but the developing forms and phases, something out of which they crystallise at the various onward stages of its progress. And if there is this more fundamental principle, can it be formulated into a definite concept, and will it account for the specific concrete character of our universe? That is our problem, in the consideration of which a commencement will be made in this chapter.

Throughout the history of human thought there have been two ultimate points of departure in the explanation of the universe, two contrasted fundamental mental attitudes or view-points from which the universe has been envisaged and accounted for. According to the one view everything is, in one way or another, given at the beginning; according to the other a minimum is assumed at the beginning, and the universe is a progressive creation or evolution from this minimum starting-point. On the first view it makes no difference whether the original creation was complete in all details or whether merely its logical or metaphysical scheme was complete, while the contents were only implicitly given. In either case there can be nothing new in the course of the subsequent history of the world. If the original creation was complete and final, all subsequent events and changes can only be rearrangements, reshufflings of the original groupings: both the material elements and their principles or forms of arrangement are there as original data, and determine all subsequent events and arrangements. If, again, the metaphysical scheme or structure of the universe must be taken as given, the evolution of the universe is merely a logical development in compliance with this scheme; or in other words, the logical development of the scheme will give us the material universe as a result. The development of Hegel's Idea is just such an attempt at a logical unfolding of the universe. In both cases the explanation of the universe is in the past, at the beginning: that beginning rules all and predetermines all. The past is the efficient cause of the future, and no new creation, nothing essentially new, can arise in the future. The full volume of reality was there at the beginning and continues to roll on, changing its forms and appearances by the way, but making no fresh addition to the original current. All real novelty and initiative, all real freedom of choice and development disappear from the universe. The process of the world becomes at most an explication, an unfolding of what was implicitly given, and not a creative evolution of new forms. This view-point has been dominant in Western science and philosophy from its early beginnings until quite recently. In physical science it fits in naturally with the orthodox laws of conservation, which preclude either the creation or the destruction of energy, mass or momentum. And in proportion as mechanistic ideas have prevailed in science and philosophy, all change has come to mean merely mechanical re-arrangements without any substantial addition to or subtraction from the sum total of reality. Those thinkers, again, who (like Leibniz) did not subscribe to the mechanistic formula were led by their theological standpoints to look upon reality as completed in the past, and to leave to the future the merely subordinate rôle of unfolding, evolving, explicating what was virtually contained in that past. This, therefore, is the very limited sense in which the term development or evolution as used by them must be understood. Where they believed in a dynamic progressive universe, they meant merely a universe which was progressively unfolding what was implicitly contained in the past. The view-point of Evolution as creative, of a real progressive creation still going forward in the universe instead of having been completed in the past, of the sum of reality not as constant but as progressively increasing in the course of evolution, is a new departure of the nineteenth century, and it is perhaps one of the most significant departures in the whole range of human thought. Not only has the old static view of reality with its fixed elements and species disappeared, the new dynamic view of Evolution does not merely negate the old static view, it has gone much further. Evolution is not merely a process of change, of regrouping of the old into new forms; it is creative, its new forms are not merely fashioned out of the old materials; it creates both new materials and new forms from the synthesis of the new with the old materials. The creativeness of matter is, as we saw, confined to the aspect of structure and to the refashioning of new structures out of the pre-existing material units: in that sense matter has only a limited though real creativeness. When we come to organisms we find a very much larger measure of creativeness in Evolution. For it is admitted that the new qualities or characters which give rise to new varieties or species are really new in the sense that they have not been there before and are not mere reshufflings of characters which were there before. New characters are created, and on the basis of them new varietal or specific forms of a stable kind arise. A still larger measure of creativeness applies to mind both in its intellectual and in its ethical aspects; thought is creative in all its activities from the simplest sensation up to the most complex judgment; and the ethical or practical reason is creative of values, moral, spiritual and religious values, in the fullest sense. Hence arises the view of Evolution as creative of the new, as an epigenesis instead of an explication, as displaying novelty and initiative, as opening up new paths and rendering possible new choices in the forward march, as creating freedom for the future and in a very real sense breaking the bondage of the past and its fixed predeterminations.

The view-point of creative Evolution is to-day embraced by scientists and philosophers generally, and this consensus between them in a matter of cardinal importance constitutes a most promising situation for the future, and may lead to far more fruitful co-operation between science and philosophy than we have known for a hundred years. Let me point to one important direction in which this co-operation is called for.

In their actual procedure philosophers have occupied themselves with general principles, while scientists, except in the domain of Mathematics pure and applied, mostly occupy themselves with the investigation of concrete things, bodies, organisms and the like. Scientists have more and more buried themselves in details, exploring facts to their minutest details, and looking to ever greater specialisation to give the clues to the unsolved problems. More than ever before they are occupying themselves with the problems of structure, the structure of matter and the physical universe, the structure of cells and organisms as explaining the systems of life, the structure of the nervous system and the brain with a view to understanding the movement of Evolution in its higher reaches. While science is thus preoccupied with the details of structure, philosophy continues very much on the old lines of exploring general points of view, general principles and tendencies and concepts. Philosophy, in endeavouring to demarcate a province of her own and distinct from the special regions ruled by science, is more and more confining herself to general viewpoints and ultimate concepts and principles, and thus runs the risk of getting further away from science instead of drawing closer to it. The result of this divorce is lamentable in the extreme. For science, divorced from the viewpoints and principles which philosophy embraces, structure becomes merely mechanism. For philosophy, divorced from the actual concrete structural facts which science studies, the general principles remain in the air, and never generate this specific concrete sensible world which is there to explain and understand. But the real world is neither a mere principle nor a mere structure, neither a disembodied soul nor a soulless mechanism. The creative Evolution which both scientist and philosopher embrace works as a general principle or tendency in and through particular concrete specific forms. Evolution is thus structure *plus* principle, interpenetrating each other, reacting on and vitalising each other. Individuation and universality are equally characteristic of Evolution. The universal realises itself, not in idle self-contemplation, not in isolation from the actual, but in and through individual bodies, in particular things and facts. The temple of the Spirit is the structure of matter; the universal dwells in the concrete particular; neither is real nor true apart from the other. All this sounds like truisms and platitudes. But if there is any force in this position it comes to this, that the pursuit of the separate paths of science and philosophy will not bring us to our goal. Their roads must be made to converge. Concepts must be developed which will include the material and the view-points of both science and philosophy. The pathway of the real is neither abstract general principles nor the wilderness of details; and if we wish to understand Evolution, we must develop concepts adequate to its actual process, concepts which will be representative of its real characters of concreteness and universality. In other words, we must construct a conceptual model which will as accurately as possible reproduce what actually goes on in the process of cosmic Evolution, our main concern being to make our explanation of Nature's process as true to actual observed facts as possible. Abstract principles alone cannot carry us to the understanding of the concrete procedure of Evolution. Structures by themselves, again, cannot generalise themselves into a universal process such as Evolution. Mere structure is not enough, because it misses the generic, the universal in reality. General principles or tendencies are not enough, because they are not concrete such as natural reality is. The two must be blended in a new concept. And it may be found that the new concept is actually not a blend of them, but the original unity from which they have been dissociated, and that the synthesis produces more than a mere concept, reveals in fact an operative causal principle of fundamental significance.

To illustrate how philosophers operate with general principles or tendencies which refuse to produce particularity, and therefore fail to explain the concrete character of reality, let us glance for a moment at Bergson's system. Any other would have served perhaps equally well, but Bergson has the great merit of being the most influential and brilliant exponent of the philosophy of Evolution in our time, and a reference to his work will therefore keep us close to our own subject matter. Bergson singles out the principle of Duration as both ultimate and all-embracing and as thus capable of both generating and explaining reality. He reaches the concept of Duration by going back into the depths of subjective experience until he comes to the point where we feel ourselves most intimately within our own life. He divests this experience of all elements of change or differentiated features, all subjective and objective items of experience are eliminated, and there remains the bare flow or passage of the inner life. This is Duration; this homogeneous flow or passage is for him the creative principle in the universe. It underlies and generates our idea of time. But time is not pure Duration, it has become infected with the Intellect, and instead of being a continuous enduring process it has become a summation of units, or points, or small unit lengths of happening. In other words, time has become spatialised, analytical and arithmetical in character in proportion as it has become divorced from its original pure form as uniform flow or process or Duration. This creative Duration is the tap-root not only of time, but also of Evolution, and is the source of all the multiplicity of forms and activities which we see in the universe. Bergson's *Creative Evolution* is an attempt to show how this fundamental principle beginning as nothing but bare flow or passage builds up the concrete universe within us and around us. To him belongs the signal merit of having elucidated the concept of Time more clearly than has been done before in philosophy. The theory of Evolution has made the aspect of time in the universe more important than ever before, and Bergson has rescued the concept of time from the confusions in which it had become entangled, not only in our empirical experience, but even in our scientific and philosophic ideas. But while freely conceding this great merit to Bergson I must confess that I fail to see how from pure Duration he has produced concrete reality. It simply cannot be done. From bare, undifferentiated, homogeneous unity you cannot reach out to multiplicity. You may call pure Duration creative, but it will create nothing until it is mixed with something very different from itself. And indeed Bergson has had to summon to his rescue another principle, which he has invested with all the characters of which he had so carefully deprived Duration. This is the

Intellect. The Intellect is practical, differential, analytical, selective, purposive; it is at once the principle and the instrument of action; it can analyse the material before it and choose what is useful for its purpose. It is spatial, it converts pure Duration into impure Time. Nor is Time its only offspring. From its marriage with pure Duration the Intellect has produced all the sensible forms in the universe. Bodies and things are simply our lines of action on matter. The practical Intellect selects what it wants for action and ignores or simply does not notice the rest. The sensible qualities it distinguishes are those which attract its attention by their practical usefulness or serviceableness for its purposes. The forms of bodies and things are therefore merely the result of this selective action on matter.[1] Structure is the creature of Intellect; in fact in its forms the sensible universe is an intellect-made universe. The result is a hopelessly lop-sided affair, and but for Instinct and Intuition—another twin of this marriage who partake more of the pure and gentle character of the maternal Duration—reality and truth in the universe would be distorted beyond recognition. Such is a very summary statement of the essential point in Bergsonism, at least as I understand it. For my present purpose I have only two criticisms to make. In the first place, as already indicated, the principle of pure Duration fails either to generate or to explain impure concrete reality. In the second place, there is far more in structure than the mere creation of the Intellect. Admitting the practical instrumental selective character of the Intellect, it would yet be a profound mistake to make it the sole cause of the forms and structures of sensible bodies or things. The Intellect is not creative in that fundamental sense. To make of Intellect what Kant made of Space and Time—the forms imposed *ab extra* on sensible reality by the activity of the mind—would be a travesty of psychology and nature alike. Intellect selects and orders, but not arbitrarily; it is itself merely an element in a greater, more universal order. Structure is the creature of experience, and experience is an interaction

[1] *Creative Evolution*, p. 102.

of the subjective and objective factors so intimate and unanalysable that it is impossible to say how much of the result is due to one factor and how much to the other. Structure is as much objective as subjective in its psychological origin. To put the forms, structures, the order of Nature to the sole account of the Intellect or subjective factor in experience is most seriously misleading and is subjective Idealism in its most dangerous form. As we have seen in the preceding chapters, structure or something in the nature of structure is inherent in the objective order of Nature, just as we shall hereafter see that it is inherent in the orders of life and mind.

Where Bergson seems to me to have gone wrong is his impoverishment of the creative principle by reducing it to the bare empty form of Duration. In order after that false step to set his Creation going it was inevitable that another mistake should be made, and that a relatively subordinate factor, like the Intellect, should be overloaded with importance. Thus the Intellect, which is a sort of Machiavelli or Mephistopheles in the Bergsonian system, has a rôle assigned to it which is accentuated both unduly and in a one-sided manner. In order to understand Nature we have to proceed more modestly and in closer touch with our ordinary observation of her ways.

Let me try to make my point clear by stating it in another way. I wish to get as near as possible to what one might call Nature's point of view in our explanation of her. To understand Nature we must take one of her own units, and not an abstract one of our own making. We must as it were take a small sample section of Nature which will include as one and indivisible both the element of activity or principle and the element of structure or concreteness in her. Our concept must correspond to such a section as our starting-point, and we must then proceed to apply it as a sort of standard with which to measure up the whole range of Evolution. In this way we shall try to explain Nature by reference to herself and her own standards, so to say, instead of by reference to intellectual abstractions of our own devising.

It may be objected that in taking such a small section or unit of Nature as our starting-point I am implicitly assuming all that follows; that I am taking a small section of the evolved in order to explain Evolution,[1] and am therefore begging the question, and that I shall be only finding hereafter what I have posited at the beginning. This, however, is not so. The criticism would have force if Evolution were merely explicative and not creative; and if my natural unit would by mere unfolding produce all the rest in the course of time. We have, however, seen that Evolution is creative; the evolution of an assumed unit factor would therefore by no means unfold the implicit contents of that factor, but would proceed creatively, and would in the end far transcend the elementary unit which was the starting-point. Let us therefore proceed in the way I propose and try to reach a concept of Nature and her progress which will not be imposed on her from without, but which will keep as close as possible to her own natural units, structures or standards, so far as we have experience of them.

At this stage we return to the difficult question which was raised at the beginning of this chapter. We are trying to dig down to the very roots of reality and to raise an issue the solving of which will be no light task. The issue has indeed become inevitable as the result of the preceding chapters.

In Chapters II and III we found the physical properties of matter were geometrically, that is in a sense mentally, determinable. We found also that matter, instead of the inertness, fixity and conservatism traditionally associated with it, was in reality plastic, mobile and transmutable in its types, and in a sense creative of its forms and values.

How have we to understand this? Is life or mind implicit in matter, and are the characters just referred to an appeal of the human mind to the mind imprisoned in matter? Has Science gone so far in her long search for truth that at last mind greets mind in the inner nature of things? Have

[1] Bergson's criticism of Herbert Spencer; see *Creative Evolution*, p. xiv.

the rescuers reached the imprisoned in the long dark tunnel of Nature?

Again, in Chapter IV we found in the organism and even in the cell a perfectly adjusted system of co-operation so closely approaching the social in character, a complicated system of controls so closely approaching the mental in character, as once more to raise the question of mind on a really extensive scale implicit in Nature. As we find life on the one hand encroaching on the domain of matter, so again we find mind encroaching far beyond its own proper domain on that usually assigned to life. Is life implicit mind, mind asleep and almost waking? Is life latent in matter, and is Mind latent in life?

What is the answer to these questions, and how have we to conceive matter, life and mind to explain this overflow into each other's domain? Is it possible to have a concept which will embrace all these facts as phases of its own creative development? Is it possible to develop the concept of a principle which is successively physical, biological and mental in its developing phases, in other words, of which matter, life and mind are the manifestation? Is it possible to have a fundamental concept of Evolution, of which matter, life and mind would be the successive stages?

This is the sort of question which naturally arises as a result of the point which we have reached in our discussion. And the answer which one ventures to bring forward must have reference not only to fundamental principles, but also to that requisite of concrete character which we have just now seen to be essential in any solution which professes to be true to nature.

The last two chapters have not only raised the question but prepared the way for the answer which will be given in the sequel. We there saw that reality is not diffuse and dispersive; on the contrary, it is aggregative, ordered, structural. Both matter and life consist, in the atom and the cell, of unit structures whose ordered grouping produces the natural wholes which we call bodies or organisms. This character or feature of "wholeness" which we found in the case of matter and life has a far more general application and points to something fundamental in the universe, fundamental in the sense that it is practically universal, that it is a real operative factor, and that its shaping influence is felt ever more deeply and widely with the advance of Evolution. Holism is the term here coined (from ὅλος = whole) to designate this fundamental factor operative towards the making or creation of wholes in the universe. Let us first try to get some general idea of what Holism is and what "wholes" are; thereafter I shall try to define these terms more closely.

We are all familiar in the domain of life with what is here called wholes. Every organism, every plant or animal, is a whole, with a certain internal organisation and a measure of self-direction, and an individual specific character of its own. This is true of the lowest micro-organism no less than of the most highly developed and complex human personality. What is not generally recognised is that the conception of wholes covers a much wider field than that of life, that its beginnings are traceable already in the inorganic order of Nature, and that beyond the ordinary domain of biology it applies in a sense to human associations like the State, and to the creations of the human spirit in all its greatest and most significant activities. Not only are plants and animals wholes, but in a certain limited sense the natural collocations of matter in the universe are wholes; atoms, molecules and chemical compounds are limited wholes; while in another closely related sense human characters, works of art and the great ideals of the higher life are or partake of the character of wholes. In popular use the word "whole" is often made to cover some of these higher creations. A poem or a picture, for instance, is praised because it is a "whole," because it is not a mere artificial construction, but an organic whole, in which all the parts appear in a subtle indefinable way to subserve and carry out the main purpose or idea. Artistic creations are, in fact, mainly judged and appraised by the extent to which they realise the character of wholes. But there is much more in the term "whole" than is covered by its popular use. In the view here presented "wholes" are basic to the character of the universe, and Holism, as the operative factor in the evolution of wholes, is the ultimate principle of the universe.

The creation of wholes, and ever more highly organised wholes, and of wholeness generally as characteristic of existence, is an inherent character of the universe. There is not a mere vague indefinite creative energy or tendency at work in the world. This energy or tendency has specific characters, the most fundamental of which is whole-making. And the progressive development of the resulting wholes at all stages—from the most inchoate, imperfect, inorganic wholes to the most highly developed and organised—is what we call Evolution. The whole-making, holistic tendency, or Holism, operating in and through particular wholes, is seen at all stages of existence, and is by no means confined to the biological domain to which science has hitherto restricted it. With its roots in the inorganic, this universal tendency attains clear expression in the organic biological world, and reaches its highest expressions and results on the mental and spiritual planes of existence. Wholes of various grades are the real units of Nature. Wholeness is the most characteristic expression of the nature of the universe in its forward movement in time. It marks the line of evolutionary progress. And Holism is the inner driving force behind that progress.

It is evident that if this view is correct, very important results must follow for our conceptions of knowledge and life. Wholes are not mere artificial constructions of thought, they point to something real in the universe; and Holism as the creative principle behind them is a real *vera causa*. It is the motive force behind Evolution. We thus have behind Evolution not a mere vague and indefinable creative impulse or *élan vital,* the bare idea of passage or duration without any quality or character, and to which no value or character could be attached, but something quite definite. Holism is a specific tendency, with a definite character, and creative of all characters in the universe, and thus fruitful of results and explanations in regard to the whole course of cosmic development.

It is possible that some may think I have pressed the claims of Holism and the whole too far; that they are not real operative factors, but only useful methodological concepts or categories of research and explanation. There is no doubt that the whole is a useful and powerful concept under which to range the phenomena of life especially. But to my mind there is clearly something more in the idea. The whole as a real character is writ large on the face of Nature. It is dominant in biology; it is everywhere noticeable in the higher mental and spiritual developments; and science, if it had not been so largely analytical and mechanical, would long ago have seen and read it in inorganic nature also. The whole as an operative factor requires careful exploration. That there are wholes in Nature seems to me incontestable. That they cover a very much wider field than is generally thought and are of fundamental significance is the view here presented. But the idea of the whole is one of the neglected matters of science and to a large extent of philosophy also. It is curious that, while the general view-point of philosophy is necessarily largely holistic, it has never made real use of the idea of the whole. The idea runs indeed as a thread all through philosophy, but mostly in a vague intangible way. The only definite application of the idea has been made by the Absolutists, who have applied the expression of "the whole" to the all of existence, to the cosmic whole, to the *tout ensemble* of the universe, considered as a unity or a being. This particular use of the idea does not interest us at this stage of this inquiry. The great whole may be the ultimate terminus, but it is not the line which we are following. It is the small natural centres of wholeness which we are going to study, and the principle of which they are the expression. And I should have thought that the matter would be of profound interest to philosophers and scientists alike. But no real use has been made of this great concept even by philosophers, while by scientists it has been steadily neglected or ignored under the iron rule of the mechanistic régime. And yet the stone rejected by the builders may become the corner-stone of the building.

Let us now proceed to consider the idea of a whole more closely; and let us once more begin with natural biological wholes, such as plants or animals. An organism, like a plant or animal, is a natural whole. It is self-acting and self-moving. Its principle of movement or action is not external to itself but internal. It is not actuated or moved by some external principle or force, like a machine or an artificial construction. The source of its activity is internal and of a piece with itself, is indeed itself. It consists of parts, but its parts are not merely put together. Their togetherness is not mechanical, but rests on a different basis. The organism consists of parts, but it is more than the sum of its parts, and if these parts are taken to pieces the organism is destroyed and cannot be reconstituted by again putting together the severed parts. These parts are in active relations to each other, which vary with the parts and the organisms; but in no case is there anything inactive or inert about the relations of these parts to each other or to the whole organism. The organism further has the power of maintaining itself by taking in other parts, such as food, but again, as we saw in the last chapter, it does so not by mere mechanical addition, but by a complete transformation, assimilation and appropriation into its own peculiar system of the material so taken in. Moreover, the organism is creative in that it is capable, under certain conditions, of reproducing itself in closely similar wholes.

This rough summary is sufficient to indicate the main general characters of biological wholes. When we reach the more advanced levels of development in the higher animals and man, we are confronted with additional characters of a psychological nature, such as intelligence, will, consciousness, central control and direction of a more or less voluntary and deliberate kind. For our present purpose of a preliminary survey in this chapter we need not consider these characters more closely. But it is necessary that we should form a clearer conception of the differences which distinguish a whole in the above sense from something which is not a whole.

In the first place, I wish to emphasise that a whole according to the view here presented is not simple, but composite and consists of parts. Natural wholes such as organisms are not simple but complex or composite, consisting of many parts in active relation and interaction of one kind or another, and the parts may be themselves lesser wholes, such as cells in an organism. Wholes are composites and not simples. The idea of a whole as a simple unique individual entity is a metaphysical view which we have to guard against. Philosophy has elaborated the concept of a unique whole which is really an absolute, indestructible and unchangeable. Plato in the *Phædo*, for instance, presented the human soul as such a whole, and from its indivisibility derived an argument in favour of its immortality. What is simple, indivisible and ultimate must necessarily also be indestructible. Natural wholes according to my view, however, are not such simple indivisible entities, which are really philosophic abstractions.

Then, again, the philosophic conception leaves no room for change, movement or development of a whole. The whole or absolute of philosophy is necessarily static. The simple unique ultimate whole cannot change or develop. It is what it is unchangeably. It negatives the idea of Evolution which is essential to the conception of wholes as here presented. The view of the universe as a whole or an absolute in the philosophic sense leaves no room for progress or development, and is in conflict with all the teachings of experience and all the most significant results of science. The parts indeed may move and change, their relations *inter se* may show a flux to which the name of development may be given. But it will not be real creative development. The absolute whole of philosophy is immutable, withdrawn in itself, and unlike anything of which we have experience in this world. The idea of Evolution as creative is the very antithesis of this static absoluteness. And this idea must be decisive for us. Anything which militates against the idea of the universe as progress and creative must be discarded by us. The creative whole or Holism must not be confused with the philosophic whole or absolute.

Having warned against a philosophical misconception, let me proceed to guard against a still more dangerous scientific misconception. The mechanical view of the universe which has been, and to a large extent still is, dominant in science is in one degree or another at variance with the conception here brought forward.

The whole is not a mere mechanical system. It consists indeed of parts, but it is more than the sum of its parts, which a purely mechanical system necessarily is. The essence of a mechanical system is the absence of all inwardness, of all inner tendencies and relations and activities of the system or its parts. All action in a mechanical system is external, being either the external action of the mechanical body on some other body, or the external action of the latter on the former. And similarly when the parts of the body or system are considered, the only action of which they are capable is their external action on each other or on the body generally. There is no inwardness of action or function either on the part of the body or its parts. Such is a mechanical body, and only such bodies have been assumed to exist on the mechanistic hypothesis. A whole, which is *more* than the sum of its parts, has something internal, some inwardness of structure and function, some specific inner relations, some internality of character or nature, which constitutes that *more*. And it is for us in this inquiry to try to elucidate what that *more* is. The point to grasp at this stage is that, while the mechanical theory assumes only external action as alone capable of mathematical treatment, and banishes all inner action, relation or function, the theory of the whole, on the contrary, is based on the assumption that in addition to external action between bodies, there is also an additional interior element or action of bodies which are wholes, and that this element or action is of a specific ascertainable character.

Wholes are therefore composites which have an internal structure, function or character which clearly differentiates them from mere mechanical additions or constructions, such as science assumes on the mechanical hypothesis. And this internal element which transforms a mere mechanical addition or sum into a whole shows a progressive development in Nature. Wholes are dynamic, organic, evolutionary, creative. The mere idea of creativeness should be enough to negative the purely mechanical conception of the universe.

It is very important to recognise that the whole is not something additional to the parts: it *is* the parts in a definite structural arrangement and with mutual activities that constitute the whole. The structure and the activities differ in character according to the stage of development of the whole; but the whole is just this specific structure of parts with their appropriate activities and functions. Thus water as a chemical compound is, as we have seen, a whole in a limited sense, an incipient whole, differing qualitatively from its uncompounded elements Hydrogen and Oxygen in a mere state of mixture; it is a new specific structure with new physical and chemical properties. The whole as a biological organism is an immensely more complex structure with vastly more complex activities and functions than a mere chemical compound. But it must not be conceived as something over and above its parts in their structural synthesis, including the unique activities and functions which accompany this synthesis. It is the very essence of the concept of the whole that the parts are together in a unique specific combination, in a specific internal relatedness, in a creative synthesis which differentiates it from all other forms of combination or togetherness. The combination of the elements into this structure is in a sense creative, that is to say, creative of new structure and new properties and functions. These properties and functions have themselves a creative or holistic character, as we shall see in the sequel. At the start the fact of structure is all-important in wholes, but as we ascend the scale of wholes, we see structure becoming secondary to function, we see function becoming the dominant feature of the whole, we see it as a correlation of all the activities of the structure and effecting new syntheses which are more and more of a creative character.

There is a creative activity, progress and development of wholes, and the successive phases of this creative Evolution are marked by the development of ever more complex and significant wholes. Thus there arises a progressive scale of wholes, rising from the material bodies of inorganic nature through the plant and animal kingdoms to man and the great ideal and artistic creations of the spiritual world. However much the wholes may increase in complexity and fruitful significance as we go upward, the fundamental activity which produces these results retains its specific holistic character all through. At first, according to our present knowledge, it appears only as a definite material structure of energy units, as a specific synthesis or arrangement of material parts, for instance, in a chemical compound or a crystal. We have already seen how this structure approaches in several respects the more holistic characters of life, and it may well be that the future progress of science will add greatly to our evidence on this point. But even as it is now known, the specific structure and character of the chemical compound makes it a sort of whole, quite distinct from mere physical or mechanical mixtures. As we proceed in the rise of Nature we see in plants how this specific structure, this synthesis and arrangement of parts and characters, assumes a new co-operative character—the character of groups of related activities which are all co-ordinated into intimate relations and functions so as to preserve the plant and maintain its activities as a whole. As we proceed to animals we find not only this intimate structural synthesis of parts and characters on a co-operative basis and with co-ordinated functions, but in the emergence of the central nervous system or brain we see a new element of control and direction, which transforms the whole system, makes its co-operation more complex and efficient and gives it an entirely new range of meaning and activity. When we come to the human stage we find the highest flowering of this central control in the human personality. We find a range of values and activities undreamt of at the earlier stages. And we find these values and activities themselves tending to become wholes in the higher ranges of spiritual and artistic production. The wholeness which was only structural, inchoate, partial at the beginning of the scale of Nature, here becomes to a large extent dominant and all-pervasive. Holism, which on the lower levels was working against almost insuperable obstructions and difficulties, here emerges in a sense victorious. It is as yet only a very partial victory. Even the most complete human personality and the most perfect artistic creation are still full of imperfections, and are still only an approximation to the ideal wholeness. And Holism, standing on that high level of attainment in the human, points the way to the future, and shows that in wholeness, in the creation of ever more perfect wholes, lies the inner meaning and trend of the universe. It is as if the Great Creative Spirit hath said: "Behold, I make all things whole."

The stages in which Holism expresses itself and creates wholes in the progressive phases of reality may therefore be roughly and provisionally summarised as follows:

1. Definite material structure or synthesis of parts in natural bodies but with no more internal activity known at present than that of mere physical or chemical forces or energies: *e.g.* in a chemical compound.

2. Functional structure in living bodies, where the parts in this specific synthesis become actively co-operative and function jointly for the maintenance of the body: *e.g.* in a plant.

3. This specific co-operative activity becomes co-ordinated or regulated by some marked central control

which is still mostly implicit and unconscious: *e.g.* in an animal.

4. The central control becomes conscious and culminates in Personality; at the same time it emerges in more composite holistic groups in Society.

5. In human associations this central control becomes super-individual in the State and similar group organisations.

6. Finally, there emerge the ideal wholes, or Holistic Ideals, or Absolute Values, disengaged and set free from human personality, operating as creative factors on their own account in the upbuilding of a spiritual world. Such are the Ideals of Truth, Beauty and Goodness, which lay the foundations of a new order in the universe.

Through all these stages we see the ever-deepening nature of the Whole as a specific structural synthesis of parts with inner activities of its own which co-operate and function in harmony, either naturally or instinctively or consciously. The parts so co-operate and co-function towards a definite inherent inner end of purpose that together they constitute and form a whole more or less of a distinctive character, with an identity and an ever-increasing measure of individuality of its own. And the whole thus formed is creative at all stages, even at the first, although this is only an inchoate, immature stage. We thus arrive at the conception of a universe which is not a collection of accidents externally put together like an artificial patchwork, but which is synthetic, structural, active, vital and creative in increasing measure all through, the progressive development of which is shaped by one unique holistic activity operative from the humblest inorganic beginnings to the most exalted creations and ideals of the human and of the universal Spirit.

We find thus a great unifying creative tendency of a specific holistic character in the universe, operating through and sustaining the forces and activities of nature and life and mind, and giving ever more of a distinctive holistic character to the universe. This creative tendency or principle we call

Holism. Holism in all its endless forms is the principle which works up the raw material or unorganised energy units of the world, utilises, assimilates and organises them, endows them with specific structure and character and individuality, and finally with personality, and creates beauty and truth and value from them. And it does all this through a definite method of whole-making, which it pursues with ever-increasing intensity from the beginning to the end, through things and plants and beasts and men. Thus it is that a scale of wholes forms the ladder of Evolution. It is through a continuous and universal process of whole-making that reality rises step by step, until from the poor empty, worthless stuff of its humble beginnings it builds the spiritual world beyond our greatest dreams.

The concept of the whole as a means of tracing the evolution of reality has several advantages. In the first place, as the whole is at once both structural and expressive of an inner general principle or tendency, its concept is as it were a working model of the natural wholes we find in the universe, and is as near as we could get to that concrete character of reality to which we should have the closest regard. The concept of Holism and the whole is as nearly as possible a replica of Nature's observed process, and its application will prevent us from appearing to run the stuff of reality into a mould alien to Nature. It will, therefore, enable us to explain Nature from herself, so to say, and by her own standards. In this way justice can be done to the concrete character of natural phenomena.

In the second place the fundamental concept of Holism will bring us nearer to that unitary or monistic conception of the universe which is the immanent ideal of all scientific and philosophic explanation. At the same time it will enable us to bridge the chasms and to resolve the antinomies which divide the concepts of matter, life and mind *inter se*. Their absolute separateness as concepts is overcome and their actual overlapping (in the way we have seen) is explained, by viewing them as phases of the development of a more fundamental activity in the universe. The concept of Holism, so to say, dissolves the heterogeneous concepts of matter, life and mind, and then recrystallises them out as polymorphous forms of itself. The monism which results is not static or barren, as monism necessarily is in the philosophy of Absolutism, but progressive, creative and pluralistic in accordance with the demands of scientific theory. We shall thus be prepared to find more of life in matter, and more of mind in life, because the hard-and-fast demarcations between them have fallen away. While accepting these terms (matter, life and mind) as generally and roughly marking off the main divisions of reality, we shall not be tempted to force their application too far, and we shall be prepared for such limits to their extensions as science may show to be necessary.

In the third place, a very real advantage will accrue from the substitution of a more definite concept for the vague and unsatisfactory idea of life. The vagueness and indefiniteness of the idea of life has proved a serious stumbling-block and has largely influenced biologists to look for the way out in the direction of mechanism. The concept of life has no definite content which makes it of any scientific value. Its value is roughly to demarcate an area from other areas; it is a name for a class of phenomena which differ generally from other classes. As such it will remain useful in Science, in addition to its popular use, which of course no amount of criticism will ever affect. The term "matter" will remain in popular use in spite of the fact that Science may completely change its meaning; its connotation may be revolutionised while it remains in use to denote a class of sensible phenomena for which there is no other equally convenient name. Similarly with the use of the term "life." It will remain useful to denote a class of phenomena, without it remaining or being useful in describing them, which will have to be done through more rigorous concepts. The concept of life is too vague to be definable and pinned down to a definite content; at the same time, and perhaps for that very reason, it is liable to be hypostatised into a substance or a force apart from the organism which it denotes. It is this abuse, in addition to its indefiniteness, which has led to its abandonment by the great majority of biologists who have preferred to see in life nothing but a specific type of mechanism. I suggest that the substitution, for scientific and philosophic purposes, of the concept of the whole for life would give far more precision to the underlying idea. Thus a definite concept, whose properties could be investigated and defined, would take the place of a vague expression, already ruined by popular use and abuse. A living organism is not an organism *plus* life, as if life were something different and additional to it; it is just the organism in its unique character as a whole, which can be closely defined. The sense in which it differs from a chemical compound considered as a whole is also capable of accurate definition; and thus it is quite unnecessary to resort to the dubious concept of mechanism in order to describe the living organism or, as I prefer to call it, the holistic organism. The concept of the whole enables us to use a technical scientific terminology, which is not vitiated by popular usage, and which is capable of accurate definition and description.

The substitution of the concept or the category of the whole for that of life will probably be found a solvent for many of the most perplexing problems in biology as well as the philosophy of life. The whole connects not only with the physical on the one side and the psychical on the other, thus maintaining the contacts of Nature; it brings to bear a perfectly definite and intelligible concept on the phenomena of "life," for which hitherto no other definite category has been found except the other misleading and misplaced one of mechanism.

In the foregoing I have tried to give some preliminary and introductory sketch of the concept of the whole, which will be further developed and filled in as this inquiry proceeds. For the sake of simplicity I have omitted reference to an important feature in that concept which I must now proceed to mention and explain. I have stated that by the whole I mean, not the All-Whole of Absolutist philosophy, but the whole as exemplified and operative in small natural centres or empirical wholes such as we observe in Nature. I must now add that by the whole I mean this whole *plus* its field, its field not as something different and additional to it, but as the continuation of it beyond the sensible contours of experience. I have before drawn attention to the vital importance of this concept, and I now proceed to explain and emphasise this point more fully.

Perhaps the most important contribution which the Theory of Relativity has made to our understanding of reality is the integration of time with our spatial conceptions of the sensible world. We are too prone to look at things merely in their spatial relations, to consider them merely as objects in space. They are just as much events in time, coming from the past, enduring through the present, and reaching out into the future. As we have seen, they are not static but dynamic in their inmost structure, they are moving and active in Space-Time; and indeed their active energy is their very essence, much more than the mere static spatial appearance which they present to the observer. As merely extended, spatial and external, objects are barren abstract concepts and not the sections of concrete reality which we know them to be. It is the time-factor that makes all the difference; Time integrated with Space is active and creative, and productive of reality. The sensible objects and things of which we are aware in Nature are active energy systems in Space-Time; they are events even more than objects and things; they are concentrated centres of happening in the physical sense just as, at a higher stage of evolution, we find minds as active concentrated centres of experience. To understand Nature properly it is essential that we should habituate ourselves to look upon material bodies or things literally as events, as centres of happening, and upon the time element in them as being no less important than the sensible space element. The limitation of objects or things to their space relations or aspects obscures and distorts their real character for us and has to be got rid of at all costs.

The effect of another serious limiting factor in our sensible experience has to be recognised and eliminated. I have already referred to Bergson's description of the intelligence as a selective, discriminative, eliminative, limiting factor in our experience of the world. But the trouble really goes deeper than that. Not only our intellect but our senses also show the same tendency and defects. All our senses are definitely limited and reveal to us directly only a limited narrow range of the properties of things. It is one of the main tasks of Science to construct instruments which will supplement the limitations of our senses. An object just visible to the naked eye presents a very different appearance when seen through a powerful microscope. The microscope, the telescope, the spectroscope, micro-photography, the X-ray spectrum with its revelation of the constitution of the atom—all these and many more are devices to extend our senses beyond their limited natural range. The combined effect of our limited sensibility and the practical selective character of our intelligence accounts in part for the fact that things appear to us limited in size and form, with definite contours and margins and surfaces beyond which they do not go and come to a dead stop. This dead stop is an illusion largely due to the defects of our natural apparatus of observation. The activities which constitute the thing go beyond the sensible contours. The material part which we popularly call the thing is merely the concentrated sensible focus which discloses itself to our limited sensibility and selective intelligence; beyond that it is the dark "field" which is formed by the activities and properties of the thing beyond its sensible focal centre. We have seen in Chapter III how the inner structure of matter results in certain physical and chemical properties which constitute its field. The field is as much an integral part of matter as the sensible part which it surrounds. Anything coming within that field will be affected by it; the field shows the same properties as the thing. The field may be viewed either as activities or as structure—as elements of force or as curves. Indeed from many points of view structure and function, curve and force are convertible terms for the purpose of describing physical effects. The essential point is that the physical field is an extension of the active energy system of the thing beyond its sensible outlines, an extension which shows the same properties and has the same effects on other things within that field as the thing itself, though with ever-diminishing force or strength as the field recedes from the thing. I have already explained how this concept of the field renders intelligible the phenomena of physical action at a distance as well as of physical causation. So far a body acts or is acted upon by external bodies this process takes place in its field and nowhere else. In their fields bodies interpenetrate each other and thus secure that continuity between them which supplies the bridge for the passage of change between them.

So much for purely physical fields. In the consideration of organisms as wholes the question of the field becomes much more important than in the case of physical bodies. What is the field of an organism? Many will be tempted to reply offhand that it is its environment. That answer will, however, be too wide and may be seriously misleading. The environment is a confused complex concept, and there is much more in it than belongs to the field of a particular organism. The field of an organism is its extension beyond its sensible limits, it is the more there is in the organism beyond these limits. To get to the field of an organism we have to answer this question: In order fully to understand the nature, functions and activities of an organism, what more is necessary to its concept beyond its sensible data? An organism appears a mystery because the sensible data are insufficient to account for its character and properties. Biologists dissect and ransack its sensible structure to find there the physical basis and explanation of its activities; but in doing so they put a weight upon that structure which is often more than it can bear. For the fact is that the sensible structure is not the whole structure, and is too narrow a base for the superstructure of organic activities which seem to grow from out of the sensible structure. For the full explanation of these activities we have to search another part of the structure which is not sensible and has on that account been ignored hitherto; I refer to the field. Biologists have tried to find in the organic structure physical elements or mechanisms to account for all the properties and functions of the organism. But there are literally not sufficient sensible elements to go round; the infinity of variations which take place in organic life vastly transcend the apparent structural organisation. A minute speck of protoplasm is supposed to carry in its structure, on a sort of point to point correspondence, the hereditary experience of the race for untold millions of years, and this structure is in addition required to account for much more besides in the individual life. The industry and ingenuity which have been displayed in this search for the inner mechanisms are above praise, but beyond a certain point the search is certain to be vain; results become mere guess-work, and the very existence of the structures and mechanisms sought for is more than problematic. The concept of the field overcomes this difficulty. According to this view the sensible structure is a narrow concentrated sensible focus beyond which is indefinitely extended an insensible structural field as the carrier of organic properties. And the question arises how we have to conceive this field and what there is in it. What has been said of the Time factor in the physical field applies with tenfold more force here. The organism much more than the physical body is an historic event, a focus of happening, a gateway through which the infinite stream of change flows ceaselessly. The sensible organism is only a point, a sort of transit station which stands for an infinite past of development, for the history and experience of untold millions of ancestors, and in a vague indefinite way for the future which will include an indefinite number of descendants. The past, the present, the future all meet in that little structural centre, that little wayside station on the infinite trail of life. But they only meet there, without its being able to contain them all. From that centre radiates off a field of ever-decreasing intensity of structure or force, which represents what has endured of that past, and what is vaguely anticipated of the future. The organism and its field is one continuous structure which, beginning with an articulated sensible central area, gradually shades off into indefiniteness. In this continuum is contained all of the past which has been conserved and still operates to influence the present and the future of the organism; in it also is contained all that the organism is and does in the present; and finally, in it is contained all that the organism vaguely points to in its own future development and that of its offspring. In other words, the organism and its field, or the organism as a "whole"—the holistic organism—contains its past and much of its future in its present. These elements are in it as active factors, the future and the past interacting with the present. The whole is there, carrying all its time with it, but clear and definite only for a small central area, and beyond that more and more fading away in respect of the dim past and the dimmer future. And this time is not the abstract time of mechanics, but real creative passage or duration in the Bergsonian sense. The biological whole is fully explained not merely in the light of its past and its present but also of its future. The force which it exerts in its field is the expression of its total time factor. It is impossible to overestimate the importance of this time factor in the development and consequently in the explanation of organism. An organism is a continuous autogenesis: behind it is its phylogeny, which it partially repeats in its individual history, and which in any case is a powerful factor in its individual development; before it, again, is the future to which it points, not only as general orientation of coming development, but more specially as the realisation of the potentialities which it holds as the seeds of the future. The pull of the future is almost as much upon it as the push of the past, and both are essential to the character, functions and activities which it displays in the present. But without the concept and the imagery of the field, which contains both the future and the past in the whole, it would be difficult to render the presence and the operation of the future as a factor in organic activity and development intelligible. The current view of structure restricts it entirely to the past and explains it as a product of the past, and therefore fails to give a complete view of it. It is unnecessary at this stage to explore other factors in the field of organism, as we shall have to make further use of the concept of the field when we come to consider the principles of organic Evolution in Chapter VIII. Enough has been said to show that in biology, perhaps even more than in physics, the concept may prove helpful in the elucidation of phenomena which it is almost impossible to explain on the narrow and confined basis of the existing organic concepts.

In explaining the important topics with which this chapter deals I do not know in how far I have succeeded in making my meaning clear. Nor do I feel sure that the ideas here developed have been presented in their best or final form. It is quite possible that in more expert hands they may prove capable of better statement and more skilful development. I trust, however, that what seems unclear and doubtful at this stage will become both intelligible and acceptable in the following chapters, where the concepts of this chapter will be further developed and applied in the explanation of organic and psychic Evolution.

Let me conclude with a word on nomenclature, intended to prevent ambiguity and misconception in the sequel. According to the view expounded in this chapter the whole in each individual case is the centre and creative source of reality. It is the real factor from which the rest in each case follows. But there is an infinity of such wholes comprising all the grades of existence in the universe; and it becomes necessary to have a general term which will include and cover all wholes as such under one concept. For this the term Holism has been coined; Holism thus comprises all wholes in the universe. It is thus both a concept and a factor: a concept as standing for all wholes, a factor because the wholes it denotes are the real factors in the universe. We speak of matter as including all particles of matter in the universe: in the same way we shall speak of

Holism as including all wholes which are the ultimate creative centres of reality in the world.

Difficulty may arise because Holism will sometimes also be used in another sense, to denote a theory of the universe. Thus while matter and spirit are taken as real or substantive factors, and material-ism and spiritual-ism or ideal-ism as concepts or theories in reference to them respectively, it would by analogy not be improper to use the term Hol-ism to express the view that the ultimate reality of the universe is neither matter nor spirit but wholes as defined in this book. And sometimes Holism will be used in that wider sense as a theory of reality. But its primary and proper use is to denote the totality of wholes which operate as real factors and give to reality its dynamic evolutionary creative character. No confusion need arise if these two distinct applications of the term are borne in mind.

CHAPTER VI

SOME FUNCTIONS AND CATEGORIES OF HOLISM

Summary.—Avoiding as far as possible philosophical categories and confining ourselves to scientific view-points, we shall now try to consider more closely the concept of the whole and the results flowing from it. We have already seen that the concept of the whole means not a general tendency but a type of structure, a scheme or framework, which, however, can only be filled with concrete details by actual experience. A whole is then a synthesis or structure of parts in which the synthesis becomes ever closer so as materially to affect the character of the functions or activities which become correspondingly more unified (or holistic). It is, however, important to realise that the whole is not some *tertium quid* over and above the parts which compose it; it *is* the parts in their intimate union, and the new reactions which result from that union. But in that union the parts themselves are more or less affected and altered towards the type represented by the union, so that the whole is evidenced in a change of parts as well as a change of resulting functions.

The whole thus appears as a marked power of regulation and co-ordination in respect of both the structure and the functioning of the parts. This is probably the most striking feature of organisms—that they involve a balanced correlation of organs and functions. All the various activities of the several parts and organs seem directed to central ends; there is thus co-operation and unified action of the organism as a whole instead of the separate mechanical activities of the parts. The whole thus becomes synonymous with unified (or holistic) action.

This intense synthesis and unification in the action of a whole involves a corresponding transformation of concepts and categories. Thus while in a mechanical aggregate each part acts as a separate cause, and the resultant activity is a sum of the component activities; in organic activity or the activity of the whole this separate action or causation disappears in a real synthesis or unity which makes the components unrecognisable in the unified result. Yet even here we must realise that the whole does not act as a separate cause, distinct from its parts, no more than it is itself something additional over and above its parts. Holism is of the parts

CHAP. VI FUNCTIONS AND CATEGORIES

and acts through the parts, but the parts in their new relation of intimate synthesis which gives them their unified action.

The whole, therefore, completely transforms the concept of Causality. When an external cause acts on a whole, the resultant effect is not merely traceable to the cause, but has become transformed in the process. The whole seems to absorb and metabolise the external stimulus and to assimilate it into its own activity; and the resultant response is no longer the passive effect of the stimulus or cause, but appears as the activity of the whole. This holistic transformation of causality takes place in all organic stimuli and responses. The cause or stimulus applied does not issue in its own passive effect, but in an active response which seems more clearly traceable to the organism or whole itself. In fact the physical category of "cause" undergoes a far-reaching change in its application to organisms or wholes generally. The whole appears as the real cause of the response, and not the external stimulus, which seems to play the quite minor rôle of a mere excitant or condition.

The most important result of the idea of the whole is, however, the appearance of the concept of Creativeness. It is the synthesis involved in the concept of the whole which is the source of creativeness in Nature. Nature is creative, Evolution is creative, just in proportion as it consists of wholes which bring about new structural groupings and syntheses. The whole involves these new structural groupings out of the old materials; and thus arises the "creativeness" of Evolution, as well as the novelty and initiative which we see in organic Nature. The concept of creativeness which flows from that of the whole has the most far-reaching effects in its application to Nature. Once we grasp firmly the fact that Nature and Evolution are really creative, we are out of the bonds of the old crude mechanical ideas, and we enter an altogether new zone of ideas and categories. But the important point for our purpose is that "creativeness" is simply a deduction from the concept of the whole and is characteristic of the order of wholes in the universe. It is wholes and wholes only that are creative. The formula *omne vivum e vivo* could therefore be generalised and applied to wholes generally. This creativeness issues not only in the origin of new organic species, but also in the great Values which are the creations of the whole on the spiritual level.

From this it is clear how also the concept of Freedom is rooted in that of the whole, organic or other. For the external causation is absorbed and transformed by the subtle metabolism of the whole into something of itself; otherness becomes selfness; the pressure of the external is transformed into the action of itself. Necessity or external determination is transformed into self-determination or Freedom. And as the series of wholes progresses the element of Freedom increases in the universe, until finally at the human stage Freedom takes conscious control of itself and begins to create the free ethical world of the spirit. Holism thus becomes basic to the entire universe of organic progress and free creative advance, to the Values and Ideals which ultimately give life all of worth it has, and to the Freedom which is the condition of all spiritual as well as organic progress.

But Holism is seen not only in the advance, in the changes and variations for ever going forward. It is seen just as much in the stability of the great Types. The new always arrives in the bosom of the pre-existing structure, and at its prompting and largely in harmony with it. Its novelty is small compared to its essential conservatism. Variation is infinitesimal compared to Heredity. It is this fundamental unity or unitariness and wholeness in organisms and organic Evolution generally which seems to explain their essential stability as well as the regulation and co-ordination of the whole process,—its conservative self-control—if one may use a metaphor.

The chapter concludes with a summary of the functions which Holism exercises in the shaping of Evolution.

IN the last chapter the ideas of the whole and Holism were sketched in a general and preliminary way. Before we proceed to test the working value of the new ideas it will be necessary to explore somewhat more deeply into them. It is the vagueness of the concept of Life which makes it unsatisfactory for scientific purposes, and we should make certain that the concept of the whole, which is intended to make it definite for scientific purposes, be made as clear as possible.

Let me here say a word about the method we are pursuing. Hitherto we have as closely as possible followed the results of science; we have studied the fundamental structures of physical and biological science in the atom and the cell, and endeavoured to frame a concept of the whole on the basis of those structures. I propose to continue to pursue this course, and to explore and build up the concept of the whole from the results of the analysis of Nature. We shall try to understand what is involved and implied in the processes of the small centres of unity in Nature, and derive as much aid and illumination from them as possible; in that way we shall try to proceed as a matter of method from the apparently simple to the complex. We are trying to build up a natural concept of the whole, and for that as well as other reasons we are avoiding a recourse to philosophical considerations. The temptation is very strong for investigators when they approach the domains of life and mind, so different apparently from that of physical science, to abandon the scientific categories of research for philosophical categories, and to seek for an explanation of the phenomena of life in concepts which sound strange and alien to science. No wonder that most biologists, frightened by this procedure and by this appeal to ideas and methods of which they are traditionally suspicious, react in the opposite direction, and seek refuge in purely mechanical ideas and explanations of the phenomena of life. At first sight the concept of the whole may appear to wear a metaphysical garb; but whatever its occasional use in other connections, the intention here is to eschew metaphysics and to hammer out a concept which will supply a real and deeply felt want in the explanation of organic processes, and which will at the same time give expression to the natural affiliations of the phenomena of life with those of matter on the one hand and of mind on the other. We shall follow the scientific clues as far as is in any way possible in the carrying out of this intention. Above all it is necessary to make the concept of the whole as simple, clear and definite as possible.

Let me repeat what was said in the last chapter: the whole is not a general principle or tendency; it is a structure or schema. A natural body or organism can be analysed into two factors: the form, structure or schema, and the concrete characters or qualities which fill up that form or structure. For these concrete characters or qualities we have in every case to rely on experience; the redness, hardness, wetness or smell of a thing or the characters of an animal can only be learnt from observation or experience in any particular case. But the form or structure involves features which can be most conveniently generalised into concepts, and if these concepts are clear and definite, results can again be deduced from them which make them most useful as counters of thought and explanation. The generalised concepts of space and time as developed, not so much by the philosophers as by the mathematicians, have these qualities of clearness and definiteness which make them specially fruitful for investigating the structure of the physical world, as we have seen in the discussion of the Theory of Relativity. And similarly the concept of the whole, if clearly apprehended and firmly held, may become a powerful means of exploring the intricate phenomena of life and mind. The concept of the whole is a generalised structure, an abstract schema, a framework to be filled in in any particular case; and it is this structural or schematic character which brings it close to that concrete character which distinguishes all natural objects in the world of experience. For the sake of clearness let us proceed to analyse the fundamental characters of a whole as we see it exemplified in, say, a simple organism.

A whole is a synthesis or unity of parts, so close that it affects the activities and interactions of those parts, impresses on them a special character, and makes them different from what they would have been in a combination devoid of such unity or synthesis. That is the fundamental element in the concept of the whole. It is a complex of parts, but so close and intimate, so unified that the characters and relations and activities of the parts are affected and changed by the synthesis. The analogy of a physical mixture and a chemical compound is very useful and instructive in this connection, and we have already seen that in a real though limited sense a chemical compound is a whole. A whole is not some *tertium quid* over and above the parts which compose it; it *is* these parts in their intimate union and the new reactions and functions which result from that union. It is a new structure of those parts, with the altered activities and functions which flow from this structure. The parts are not lost or destroyed in the new structure into which they enter; the atoms or molecules persist in the new compound, just as the cells persist in the organism. But their independent functions and activities are, just as themselves, grouped, related, correlated and unified in the structural whole. To the structural unity of the parts in the whole corresponds an equally and perhaps even more significant functional unity or correlation of activities. Just as in dynamics a body subject to pulls in various directions moves with one resultant velocity in one definite direction, so the functions and the activities of the parts in the whole are all co-ordinated and unified into one complex character which belongs or appears to belong to the whole as such. With this difference, again (just as in the difference between a physical mixture and a chemical compound), that the resultant function is not a mere addition and composition of the unaltered composing functional elements, but the change involves both the elements and their final result. Thus taking x to represent a mixture and x_1 to represent a whole, we cannot say that $a + b + c + d = x$ (mixture) in the one case and $= x_1$ (whole) in the other; but in the synthesis which results in x_1 the functions of the parts themselves are changed into a_1, b_1, c_1, d_1, so that corresponding to the formula of mixture $a + b + c + d = x$ we have the holistic formula $a_1 + b_1 + c_1 + d_1 = x_1$. It is most important to realise this point; both the individual functions of the parts (cells, organs, etc.) and their composition or correlation in the complex are affected and altered by the synthesis which is the whole. Not only does the synthesis of the parts influence and indeed constitute the whole; the whole in its turn impresses its character on each individual part, which feels its influence in the most real and intimate manner. The whole-ward tendency and activity of the parts is most deeply characteristic of the nature of the whole. This, then, is the primary and most important element in the concept of the whole: the synthetic unity of structure and its functions which affects the parts and their functions or activities without their loss or destruction. The unity, although so close and intimate and so deeply affecting the parts and their functions, is not such as to merge the parts completely, but to leave them a latitude which varies with individual wholes at the same state of development, and still more at different stages of development. To this latter point we shall revert hereafter.

From this fundamental unity of the parts which constitute the whole, and the intimate reciprocal influence which parts and whole exert on each other, follow certain results of great importance for our general concept of the whole.

In the first place, unity of action, which is characteristic of the whole, shows itself in the marked power of regulation and correlation which the whole appears to possess in respect of its parts. This is perhaps the most striking feature of organic wholes; however complex they are, a certain balanced correlation of functions is maintained. If there is any disturbance among the parts which upsets the routine of the whole, then either this disturbance is eliminated by the co-operative effort of many or all the parts, or the functions of the other parts are so readjusted that a new balance and routine is established. The synthetic unity of the whole produces synthetic or holistic action throughout the whole; the activities and functions of the parts also become holistic, so that in addition to their ordinary routine they have a whole-ward aspect or tendency which becomes active whenever the balance of the whole is disturbed. It is this holistic character distinguishing the activity and functions not only of the whole but also of its parts which underlies the remarkable phenomena of co-operation among cells to which attention was drawn in the fourth chapter. The co-operation is not so much the interaction of independent units as in truth and really the pressure of the whole on the parts. Indeed the entire function or system of the organism is holistic; the synthetic unity of the whole is so deeply stamped on the parts and reflected in the activities of the parts, that they all appear to "play up" to each other, and to co-operate in maintaining or, in case of disturbance, restoring the balance of equilibrium of activities which is characteristic of the particular whole. From the synthetic unity of the whole follows the holistic action of all its parts, as well as the characteristic power of correlating and regulating which the whole seems to exert in respect of the parts. All these properties really flow from the idea and nature of the whole; once this idea is clearly realised, the true principle of organic explanation is found, and the application of the ideas and methods of mechanism or vitalism becomes superfluous, as we shall see later.

In order to assist us in rendering clear our ideas of the whole and holistic action, as distinguished from those of a mechanical aggregate and mechanical action, let us consider a material system in dynamic equilibrium, which has many analogies to the ideas we are exploring. The character of such a system is that within certain limits it will maintain its equilibrium against disturbance and interference. If it meets with any disturbance, such as an external impact or any interference with its internal movements, its equilibrium will for a moment be disturbed; but immediately readjustments will take place, the effects of the disturbance will become distributed throughout the system, new positions of the parts and new movements of these parts will result, with the effect that a fresh equilibrium is established, and the system, with a somewhat altered arrangement of parts and movements, will once more be in dynamic equilibrium. When we pass from this physical system to an organic whole a transformation of ideas takes place: the system becomes a synthesis qualitatively different from the system, a synthesis so intense that a new unity arises and a different order of ideas becomes necessary for its explanation. To the mechanical readjustment of the parts correspond the regulation and correlation which the organic whole exercises in respect of its parts; with this difference, again, that while the new mechanical equilibrium is the exact mathematical resultant of the component forces, in the organic whole, parts and whole reciprocally influence and alter each other instead of merely the parts making up the whole, and in the end it is practically impossible to say where the whole ends and the parts begin, so intimate is their interaction and so profound their mutual influence. In fact so intense is the union that the differentiation into parts and whole becomes in practice impossible, and the whole seems to be in each part, just as the parts are in the whole. There is an intensification of synthesis or unity, as we rise from the mechanical composition to the chemical compound, and from this again to the organic whole; an intensification which is already qualitatively different as between the first two, but which becomes entirely *sui generis* in the last. To mistake the unity of the whole for the mere mechanical system of dynamics, and the holistic action which shows itself in organic regulation, correlation and co-operative interaction for the readjustment of forces and self-maintenance of equilibrium in the dynamic system, is to confound two quite different orders of ideas and facts. The whole differs essentially from the mechanical system; the holistic action or function differs even more essentially from the "action" or "reaction" of dynamics, and from the meaning of those terms as used in Newton's Laws of Motion and still current in the physical science of to-day.

In other directions too the nature of the whole brings about this intensive transformation of concepts. Let us take the idea of cause and see how it is affected by the concept of the whole. The causal idea is quite an interesting test to apply to the whole, and it will help to elucidate the point we are dealing with, as well as some other points that concern the nature of the whole. The question, for instance, whether the whole is something different from and additional to its parts is paralleled by the question whether the whole acts as a cause as distinct from its parts; in other words, is the causality of the whole exhausted by the causal operation of its parts, or is there something over and above the influence of the parts which must be attributed to the whole as such? I have already explained that the whole is nothing but the specific synthesis of the parts and not something additional to them. Similarly the causality of the whole is not an additional factor, but simply the causality of the parts in their intimate synthesis in the whole. In mechanical composites each element in operation or action has its own effect and is a separate cause; and the final result is the resultant blending of all these separate effects. In the whole, as we have seen, there is not this individual separate action of the parts; there is a synthesis which makes the elements or parts act as one or holistically; and the action or function is an inseparable holistic unity just in proportion as the synthesis is a whole or realises the character of wholeness. It is in this sense, and in this sense only, that the whole is a cause; it is a cause not apart from its parts, but solely through their synthesis in action. The whole fuses the action of its elements into a real synthesis, into a unity which makes the result quite different from what it would have been as the separate activities of the parts. The structural synthesis of the whole results in a similar synthesis of activity or function. Just as the whole as a structural unity, and only as such, is something different, something new compared with its parts in their separateness or isolation, so too its action is radically different from the blending of their separate actions.

Thus the causality of the whole is explained, and from this explanation one can appreciate how immensely complicated the action or functioning of an organism must be. When a stimulus is applied to an organism a whole is set in motion, and the response which results is not merely an affair of the original stimulus, but of the entire whole in all its unique complication of parts and functions which has been set in motion. The comparatively simple, isolable phenomenon of causation as observed in the interaction of material bodies undergoes a complete and radical transformation when observed in the case of an organism; and the difference is not a mystery, but is deducible from the nature of the whole as exemplified by an organism. I shall return to this matter and just now pass on to another important result of the nature of a whole.

In the preceding chapters I have more than once used

the word "creative." I have called the whole "creative"; I have called Evolution "creative"; I have even applied the term to matter in its structural characterisation. It is important to see in what sense wholes, or Evolution generally, or even matter is called "creative," and the foregoing discussion will have prepared us for the explanation which follows. There is a sense in which the word "creation" is unintelligible and falls outside the scope of an intelligible science or philosophy. I refer to absolute creation—creation, that is to say, out of nothing. Absolute creation just as absolute annihilation cannot be comprehended by the human mind. *E nihilo nihil fit* is a fundamental principle of thought as well as of Nature. But there is another form of creation which is not only intelligible but follows directly from the explanation of holistic action which I have already given. Holistic action is creative, and is the only form of creation or creativeness which is intelligible to us. Here again the distinction between mere physical mixtures and chemical compounds illustrates the difference between what is and what is not creative. A mere mechanical aggregate is nothing new, and is no more than the sum of the mixed ingredients, while the chemical compound is new in the sense that out of the constituent materials another qualitatively different substance has been made. A new structure has been formed in the chemical compound. In the same way a new structure and substance is made in the atom out of the qualitatively different electrons and protons. It was on this account and in this sense that we called matter creative; creative, that is to say, of structures and substances different from their constituent elements or parts.

It is, however, when we come to consider organisms that we see the whole creative in a full and proper sense. In thought we distinguish between the deductive and the inductive—between the deduction of the particular from the general, the drawing out, unfolding, or explicating what is given, and the reverse inductive process, the integration or synthesis of the given parts or elements into a new, more complex content. The action of organisms proceeds on the analogy of induction. We have seen how the characteristic feature of organic process is metabolism, the transformation of the given materials into something quite new, of the inorganic into the organic, of the organic material of one kind into that of an entirely different kind. Creative synthesis is the inmost nature and character of all organic actions and functions.

But it is not in metabolism only that this creative transformation is exemplified; perhaps the phenomena of growth and development afford even more characteristic and significant examples of the creative synthesis which is the clearest expression of the nature of Holism. Creation is stamped on the face of organic nature; the differences which separate individuals into species, genera, orders and so forth are real differences which were either originally created in one great creative Act at the beginning, or were creatively evolved in the gradual process of organic descent, so that the Process is creative. Everywhere we meet the new, which is irreducible to the old elements from which it seems to have sprung; the qualities and characters on which new stable varieties or species are founded cannot be explained on the basis of known pre-existing qualities or characters. And even where it is possible to recognise certain of the old elements in the new, the new is something different and contains something more than the old elements. We may say that the creative synthesis consists in the making of a new arrangement of old elements, that the old elements have been fixed in a new structure which has different qualities from the old pre-existing structure; in this case it would only be the synthesis which is really new. But to my mind if we take big stretches of organic descent and compare the main great Types which distinguish the Vegetable and Animal kingdoms, we must inevitably come to the conclusion that there is more than this in the creative process of evolution. Compare a protozoon with a vertebrate animal, or one of the higher animals with man, and it surely becomes foolish to say that the elementary units are the same in both cases, and only their arrangement or synthesis into structures is different. Mere rearrangement of supposed unalterable pre-existing elements, mere reshufflings of the old cards, will give us a sort of chemistry of Evolution, but not the vast range of real effective development which we know. The process of creative Evolution is not a mechanical rearrangement of old material; it involves the qualitatively new at every stage, from the most minute elements to the most complex structure. It is not merely the structure which is new and different from what has gone before, some of the materials are also new; the details of the new structure also involve new smaller structures along with the old inherited structures; and in the final analysis (if that were possible) we would find among the elementary units also new ones in addition to the old ones carried forward in the process of descent. There is the creation of the new variety or species (the new structure); there is the creation of new unit characters (parts of the structure) which justify the new species or variety; and there is behind the new unit characters not mere rearrangements of elements of old character units, but an integration of new materials or quality elements with the old elements in the formation of the new unit characters. The process is creative of the new at every step and at every stage, and in the smallest quality elements no less than in the large or total structures. Starting from imaginary elements $a, b, c, d,$ we find in organic advance in no case a mere structural regrouping of these elements only, but everywhere an element of qualitative newness, x, incorporated along with the old elements into the new structure. In every advance we would find not merely a new structure but also an x in one form or another. I have used the concept of units or elements for the purpose of illustration; but really the creative process of Evolution is holistic, and in the last resort unanalysable into definite units; the blending of inherited with new structures and characters is so close that no dissociation is possible, and it is impossible to say where the old elements end and the new begin. But beyond any manner of doubt in the advance the new is there along with the old in such a way that we can only understand the process, as a whole and its minutest detail, as a real creative one.

It may be argued that my view of organic creativeness as meaning, not merely new grouping or structures but also new character units and quality elements, brings us back to that conception of absolute creation which I have already declared to be unintelligible. Is it not better then, it may be asked, to fall back on the idea of potentiality rather than creativeness, and to conceive the organic advance as the rendering actual what was implicit and potential in the organism in the beginning? In this way new characters which emerge in the course of organic descent would not be taken as absolutely new, but as the appearance or emergence of potential characters which were there all along in the ancestors of the new organism. My answer would be that the concept of potentiality is quite useful but not applicable here; the formation of the new is usually a very long process which may occupy an indefinite number of generations before its actual emergence in a new sensible character or species. During this process of subsensible growth or incubation the character may be fairly described as potential in the ancestors of the new species. But to go further and to say that the new is there in potential form from the very beginning is to fall back into the preformation view of Evolution which we have already in the last chapter discarded as making a farce of all real organic advance. Potentiality presupposes that the real creative work is already done, and that the slow finishing touches alone remain to be put on. We have simply to face the facts of Nature frankly as we find them; and to my mind there is no doubt, however hard it is to picture to ourselves the underlying idea of creation, that the emergence of the really new, in other words, the creativeness of the evolutionary process, is the only view which is in harmony with our scientific knowledge. Real creativeness is a fundamental characteristic of holistic structure and action. It necessarily means such integration of structures and activities as results in new characters not there before and which cannot be reduced to pre-existing elements. Holistic action, therefore, necessarily issues in real progress and creative Evolution.

There is no doubt that the concept of creativeness raises a fundamental issue in respect both of reality and knowledge. If there is this evolution or making not only of new wholes or structures, but also of new quality elements therein, the whole fabric of Mechanism as ordinarily understood is shaken to its foundations. The iron rule of the past is broken; the future is not a mere rehearsal of the past; in many cases the new effect is more than its pre-existing cause. The universe ceases to be a hide-bound, cast-iron, completely closed system from which real progress and freedom are excluded. It is open in one direction, the direction in which time is moving; the future faces an open gateway, and the universe is the highway of the creative movement, of that great march in which all the units and formations of Reality take their part and advance towards a fuller measure of Wholeness. The freedom is limited, the movement is slow, the character of the universe is essentially conservative. But at any rate conservatism is not the last word that can be said about it. It does not go like a clock, completely manufactured, and once for all wound up at the beginning to mark a time fixed and predetermined for it. It is slowly making itself, it is slowly winding itself up, it is slowly making its own time. It is a slow, tentative, perhaps in details a somewhat blundering process; but it is real and creative; the new successful effort is for ever issuing out of the old mistakes, and a slow advance is being laboriously recorded and continuously maintained. This figurative language, although perhaps somewhat highly coloured, is really no exaggeration. "Creativeness" is the key-word, and it is also the key position in the great battle which is now being fought out between the nineteenth-century and the twentieth-century conceptions of the nature and trend of the universe, between Mechanism as ordinarily understood and what is here called Holism. Those who wish to defend the old position of Mechanism (and they are still the great majority in the army of Science) will have to concentrate their forces at this point. If the concept of creativeness, of the emergent new in the Evolution of the universe, really wins through, Mechanism as a scientific and philosophical category will be reduced to very modest proportions.

The creativeness of Evolution at all stages has indeed the most profound effect on our views of Nature and her order and on all our methods of explaining her processes. To illustrate this let us revert once more to the oft-quoted difference between a mere physical mixture and a chemical compound. The mixture is, like the compound, a structure, much looser, of course, than the compound, but still a structure of sorts. But the compound differs from it in being a radically different structure, a new structure has emerged in the compound, a creative element has entered into the process which was not there before. And the result is a complete difference in all the pertinent phenomena of the two. Even our very categories of description have to undergo a corresponding transformation. If we tried to describe the properties and actions of a chemical compound on the same principles as those of a mechanical mixture, we would go grievously wrong, and the real facts would be hopelessly distorted. The creative element which has entered into the chemical compound in its passage from a mere physical mixture, the new structure which has resulted from the change, requires new concepts and principles of description. And this is freely admitted by chemists and physicists alike. The case for new categories of description becomes far stronger at the next creative advance, where chemical structure is transformed into organic structure. In both cases Holism is at work and produces a new structure; in the organic structure the creative advance is admittedly far greater than in the chemical compound. The concentration and intensification of structure which we call the whole in the organism is comparatively far greater and higher than the similar phenomenon in the chemical compound as compared to the mere mixture. Something indisputably new has been produced; there has been creation; a new structure has arisen which has its own categories of description; and to apply mere chemical and physical concepts of action and description to this new structure is to ignore the creative advance which has taken place and to confound two entirely different, however closely related, structures and stages of Evolution. The physico-chemical view and explanation of organism therefore rests on a fundamental misconception and on a denial or disregard of the creative element in natural Evolution. There is the physical description of the mixture; there is the chemical description of the compound; and there is what I call the holistic description of the organism, which recognises the qualitative newness and *sui generis* nature of the structure which bears the characters of what we call life. The apparent materials may even be the same in all three cases; but the character and intensity of their union in each case varies in such a way that entirely different structures with entirely different characters result. There is a rising element of wholeness in all three structures, and the holistic character is by no means confined to the third or organic structure. But its wholeness is much more marked and pronounced than that of the other two; it is, in fact, the very type and exemplar of a whole; and a purely physico-chemical explanation of its nature and functions cannot possibly do justice to this unique holistic character. This is but another way of affirming the creative character of the advance from the mere physical mixture to the chemical structure, and still more in the advance from the chemical to the organic structure in Nature. The creative advance is the fact, to which our conceptual theories of explanation have to conform. The creativeness consists in the progressive advance in respect of the character of wholeness which distinguishes the three stages of structure; and the advance is in a geometrical rather than an arithmetical progression, if one may be allowed a mathematical analogy. The physical and chemical categories still apply, but they are not sufficient, and have to be supplemented by the holistic categories which correspond to and express the greater and qualitatively more intense holism which characterises the organism as distinguished from the mixture or the compound.

Let me here point out that it is not all causation which is creative; much of the causation in the universe is purely mechanical and produces nothing new. Only wholes are creative; only the causality of wholes produces effects which are really new. It is conceivable, it is even possible, that out of free scattered protons and electrons new atoms, say of Hydrogen, new physical structures may be synthetically built up, just as it is conceivable and probable that organic wholes or life structures have arisen from purely inorganic materials. That would mean the creation of matter or the spontaneous generation of life—both still unrealised possibilities from an experimental point of view. Structure in fact and in Nature arises from pre-existing structures whether in the organic or inorganic domain. *Omne vivum e vivo* is a formula which applies to all wholes and not merely organic wholes. Only wholes produce wholes, and only in wholes does the new emerge; wholes form the pathway of creative reality; only the causality of wholes is creative of the new.

This is so because of what I have already pointed out above when discussing how a whole transforms a "cause" or stimulus applied to it into something quite different from what it was before. If an external "cause" is applied to an organism or living body it will become internalised and transformed, and will be experienced as a stimulus, which in its turn will be followed by a response. The response is not the mere mechanical effect of the cause, and this is due to the complete transformation which the latter has undergone. In the moment which elapses between stimulus and response a miracle is performed; a vast series of organic changes is set going of which comparatively little is known as yet. The inorganic becomes organic, the alien stuff of the environment is recreated into the stuff of the living organism. The organic changes which take place are assumed to have their physical and chemical equivalents; but even if that is so, they are much more than the mere physico-chemical tale they tell. The stimulus has been transformed and absorbed and become a series of states of the organism; the organism has made the stimulus its own, as it were. And as a result the response is not the mere passive effect of the stimulus, but is the free and spontaneous movement of the organism itself under the influence of the appropriate stimulus. The passive external stimulus has been recreated into an active, free response of the organism. Anything passing through the organic whole thereby becomes completely changed. Any action issuing from it has the stamp of the whole upon it. The procedure is transformative, synthetic, recreative, holistic, and the result is "new" in one degree or another.

From this it will be seen that if the concept of causation is to be retained in connection with organic or psychical activities, it will have to be substantially recast. The resultant activity of an organism under a stimulus is never the effect of that stimulus, as it would be in the case of mechanical action, but always of the stimulus as transformed by the organism; the organism appears as the dominant element in the causal concept, and the stimulus appears in a minor rôle. The more active the state of the organism, and the more thorough its reaction to the stimulus, the less is the influence of the stimulus on the response, which appears as the free and almost original action of the organism. The organic response is often so great compared to the stimulus, it is so out of all proportion to it and so transcends it in every way, that the organism appears clearly as the real cause, and the stimulus merely as a minor condition or excitation.

It is thus seen that organism is a new system, with its own activities and laws and categories of action and description. It is a new centre, with a large measure of independence of the environment. This does not mean that the environment does not influence it, but it means that the environment influences it only indirectly and after a more or less complete transformation and metabolism of such influence. *Vis-à-vis* the environment the organism is something new, something *sui generis,* which does not passively accept and reproduce the influence of the environment, but utilises and appropriates it for its own purposes and in its own ways, as if it were some superior arbiter and disposer of the whole situation. The concepts of dominion, of mastery, of creation which the orthodox view places at the beginning of things are now distributed and assigned to all organisms, whose inmost nature it is only possible to express through these concepts. In other words, organism is not an effect of external causes; nor are its states and characters effects of external causes; it is in a large measure its own cause—*causa sui*—and the cause of its own states and functions. External environmental influences are merely the rough material with which it works and builds up its own system. And in the act of building the material is itself more or less completely changed into the character of the structure built. I say "more or less," because this character of creative mastery and transformation which organism displays in respect of external influences and materials is itself of a progressive character. In the lower organisms there is much more of passive acceptance and response than in the higher; and the whole process of Evolution is largely a continuous growth towards organic independence and self-regulation; in other words, towards wholeness. The concept of wholeness contains and explains all the distinguishing organic attributes in their various grades throughout the wide range of organic Evolution.

Thus it is that the creative element in Evolution, the emergent new, is associated with the nature and action of wholes, and is confined to them. Not only is the activity of wholes holistic and creative, as yet it is the only creative activity of which we have knowledge.

From this discussion it is clear how the concept of Freedom is rooted in that of the organic whole. The whole is free, the parts are bound: such would be a formula of metaphysics. For beyond the whole there is nothing external to determine it, and it is therefore free; while the parts are necessarily bound by their relation to the whole and to each other in the whole. But we are not concerned with metaphysical wholes, but with those of reality, such as organisms. And we have seen how the functioning of an organic whole releases it from the domination, the causation of the external, and conduces to its freedom. The external causation, the stimulus which operates on it *ab extra*, is transformed by its subtle metabolism into something of itself; otherness becomes selfness; the pressure of the external or the other is transformed into the action of itself. The organism is largely detached from its surroundings and centres in itself. Necessity is transformed into freedom. The causal chain of physics becomes the new badge of freedom. The whole, therefore, even in its most humble organic forms, lays the foundations of the new world of freedom. We can arrive at the same result by another process of reasoning, based on the creative activity of Holism. Under the physical system the effect equals the cause, and is therefore completely determined by the cause, *Causa = effectum*. But we have seen how this formula disappears before creative Holism; how the effect comes to contain the new and therefore to transcend its cause. The element of newness, of novelty in the holistic order of the world, means a release from the complete bondage of matter and its causality and necessity. It means a certain latitude for expansion and growth. It widens the range of possibilities; the strait and narrow path of physics becomes the prospect which ultimately widens into the great horizons of life and mind. Freedom broadens out into a world of opportunities. The animal finds that it is no longer imprisoned in its cell like the plant; it begins to move about. Gradually it learns the great lessons of direction and self-direction. The great Experiment of life assumes ever-widening degrees of freedom, until finally at the human stage freedom takes conscious control of itself and begins to create the free ethical world of the Spirit. With that development we shall deal at a later stage of this work. Here it must suffice to have pointed out the humble beginnings of freedom. And even at this stage it is important to bear in mind that the domain of life is largely distinguished from that of matter and energy by its greater degrees of freedom. Scientists speak of the degrees of freedom even in an inorganic situation. And by this they mean the element of contingency which seems inseparable even from the purely physical order. The causal chain of Nature, the necessity which characterises the procession of physical events, does not exclude elements of chance or contingency. An event may happen in this way or that way; there are alternatives between which only the actual fact can decide. To these possibilities or alternatives the phrase "degrees of freedom" is applied. But in the domain of life it acquires an added meaning, or rather, let me say, a real meaning. Life is not entirely bound, even in its most primitive forms. Hence its trials and experiments, its variations, its novelties and its creativeness, which become ever more accentuated in its progress. Evolution traces the grand line of escape from the prison of matter to the full freedom of the Spirit. It is clear that the beginnings of freedom are laid far back in the early dawn of life itself, if not earlier.

The above discussion of unified organic functioning, or unity of action, causation, creation and freedom will suffice to indicate how the whole as factor and concept involves a transformation of physical actions and categories. They have been selected as samples of holistic functions and categories and are not intended to be in any sense an exhaustive enumeration of such functions and categories. A full list would, for instance, include "individuality" and "purposiveness" as essential features, functions and categories of wholes. They are, however, referred to in Chapter IX in connection with Mind as an organ of Holism. At this stage it is only necessary to make the briefest reference to them. Thus with regard to the character and category of individuality it is only necessary to point out here that individuality is distinctive of wholes. Wholes are not arbitrarily divisible and the divided parts are not arbitrarily interchangeable. Every whole has a real character, a unique identity and an irreversible orientation which distinguishes it from everything else and is the very essence of wholeness. And this character of individuality rises with the rise of wholes in the scale of Evolution, and acquires decisive importance at the ultimate level of human Personality. Purposiveness, again, is a special form of that unified organic action which has already been discussed. It means a correlation and unification of actions towards an end, whether this is consciously conceived or apprehended or not. On the animal plane and especially on the psychical level of Evolution it is quite distinctive of wholes. In an exhaustive treatment of holistic characters and categories individuality and purposiveness would have no less important a place than those above discussed.

Let us now pass on to consider organism as a centre of internal regulation, adjustment and co-ordination of its own functions and activities. The phenomena that meet us here are indeed most wonderful. No cunningly devised machine of human contrivance can rival or even approach in delicacy of co-ordination or fineness as well as complexity of adjustment the organic wholes we see in Nature. Professor Haldane has described the wonderful combination of processes which go to make up the physiology of breathing [1] —a combination which is marvellous enough under normal conditions, but which becomes far more so when we see how curiously breathing adjusts itself to abnormal conditions, to situations artificially brought about, which it has probably never had to face in all time. No "experience" or hereditary "memory" can guide it here; and yet it rises to the occasion every time, within a wonderfully wide range of adaptability and plasticity. Practically every major physiological function shows the same power of co-ordination of

[1] *Organism and Environment*, 1917.

various organs and activities, and the same delicacy and ingenuity of adjustment to novel situations. Any one of the functional elements involved would be a wonder in itself; but when the co-ordinated combination of all is studied; when, moreover, the great variety of adjustments of this combination to unusual situations is considered, the marvel becomes baffling to our human intelligence. The most delicate processes, involving vast numbers of co-operating factors, happen not clumsily or slowly, but most finely and as it were in less than the twinkling of an eye.

What guides and controls such a complex physiological process? Intelligence such as we know it is clearly not equal to the task; nor have we any reason to ascribe intelligence to organic processes. The assumption of a vital force explains nothing, as our problem concerns something far more subtle and directive than force. Again, to look upon it as a marvellous self-working mechanism does not meet the real situation, which is more than one of mechanism, however marvellous. The theory of Evolution presupposes an original start from simple beginnings, which have multiplied, evolved and become complex in the course of Evolution. The pure chance presupposed by Mechanists has never ruled the world. There has never been a blind sorting out of possibilities according to the laws of probability; and if there had been, the chances against the present organic situation in the world would have been infinite. Not thus has the new arisen and gone forward. The new has always arisen in the bosom of the old, and under its ægis and influence. Not blind chance or contingency but the existing state of affairs has always shaped the course and direction of Evolution. The new arises from the old and largely at its prompting, and thus in harmony with it. Its novelty is very small compared to its essential conservatism. Variation is infinitesimal compared to Heredity. It is this fundamental character of unity, unitariness and wholeness as distinct from mechanical aggregation of parts, which seems to me to explain the phenomena of organic regulation and co-ordination. Organisms, of course, contain a great deal of mechanism; the detailed processes and functions are largely carried out by what one might call organic mechanisms, structures, with particular functions assigned to them. But the unification of the whole system and its self-regulating character; its plasticity of co-ordination and adjustment under all the situations of the environment which it has to face and to which it has to adapt itself; its creative movement in time, so different from what one would expect on the second law of thermodynamics; the unique facts of growth, restitution and reproduction, which not even a strained application of the mechanistic scheme would fit—these are facts and features which Holism alone can properly justify and explain.

It may be convenient if, before concluding this chapter, I give a summary of the various functions which are here assigned to Holism in the shaping of reality.

(A) 1. In the first place, Holism is a creative factor, and as such shows itself in the upbuilding and differentiation of organic structures and their functions. These may be modifications or variations or mutations. They may be ordinary specific differences such as explain the origin of different species. These differences may include new organs and structures, or merely the general complexifying of existing structures which makes organisms as a whole more complex. All these aspects of Holism are discussed in Chapter VIII.

2. This creative Holism is, of course, responsible for the whole course of Evolution, inorganic as well as organic. All the great main types of existence are therefore due to it, such as the atom, molecule, cell, organism, the great groups of plant types, the great groups of animal types, and finally the human type. Creative Holism is thus responsible for all the great divisions of Science. The cursory discussion of these aspects of Holism is spread over various chapters.

(B) 1. In the second place, apart from the detailed structural and functional differentiations above referred to, Holism is a general organising, co-ordinating or regulating factor in organisms over which it exercises a measure of guidance, direction and control. The nature of this regulative or controlling activity is discussed in this chapter, and the difficulties it gives rise to in its relation to the body or the energy system generally are discussed in Chapters VII and X.

2. This regulation and control is exercised over the structures and functions of organisms generally, but sometimes special holistic organs are evolved, which seem specially destined to assist in the exercise of this regulation and control. Such special holistic organs are the ductless glands which pour regulative secretions into the general system, the nervous system, and especially the brain with its correlate mind. These and other holistic organs are special aids to Holism in its regulative activity. Various aspects of these holistic organs are discussed in different chapters.

(C) In the third place, in order to express and explain these activities of Holism at the different grades of Evolution and at the various levels of differentiation of types and structures, categories of the Whole or holistic categories are necessary, some of which have been discussed in this chapter and others are dealt with in other chapters. Thus arise the physical, chemical, organic, psychical and personal categories, which are all expressive of holistic activity at its various levels and reducible to terms of Holism.

Holism thus appears in this scheme as the fundamental activity of the universe from which all others are derived; and the concept of Holism is the ultimate category of description and explanation from which likewise all other categories are derived. Holism therefore constitutes the ultimate view-point from which to orientate our survey of all the various forms and departments of reality.

(D) There is one more aspect of creative Holism which I must for the sake of completeness mention, although its exposition falls outside the plan of this work. We have seen that Holism is creative of all structures, inorganic as well as organic. Thus all the types of structure in the worlds of matter and life are its work. But more; as we proceed upward in the course of Evolution we find Holism the source of all values. Love, Beauty, Goodness, Truth: they are all of the whole; the whole is their source, and in the whole alone they find their last satisfying explanation. Holism not only prescribes the law in the world of structures, forms and organisms; it is the very ground and principle of the ideal world of the spirit. It is in the sphere of spiritual values that Holism finds its clearest embodiment in fact, and its most decisive vindication as an ultimate category of explanation. Its creativeness will nowhere be found more fruitful than in that last and highest reach of its evolution. Here it would be premature to do more than merely refer to this aspect of creative Holism. The exposition of its creative activity in shaping the great Ideals of the Whole is, however, too large a task to be undertaken in this introductory work.

CHAPTER VII

MECHANISM AND HOLISM

Summary.—The discussion in the last two chapters has disclosed a grading-up of such structures as can in any way be called holistic; beginning with the physico-chemical structures, into which physical and chemical relations enter; passing on to bio-chemical structures or organisms, into which those relations plus something new, usually called life, enter; and culminating in psycho-physical structures, in which all three relations enter, together with the new elements of mind and personality. In this grading-up the earlier structures are not destroyed but become the basis of later, more evolved synthetic holistic structures; the character of wholeness increases with the series and the elements of newness, variation and creativeness become more marked.

Mechanism is a type of structure where the working parts maintain their identity and produce their effects individually, so that the activity of the structure is, at least theoretically, the mathematical result of the individual activities of the parts. With the two concepts of Mechanism and Holism before us we can see how the natural wholes of the universe fall under both concepts. There is a measure of Mechanism everywhere, and there is a measure of Holism everywhere; but the Holism gains on the Mechanism in the course of Evolution, it becomes more and more as Mechanism becomes less and less with the advance. Holism is the more fundamental activity, and we may therefore say that Mechanism is an earlier, cruder form of Holism; the more Holism there is in structure, the less there is of the mechanistic character, until finally in Mind and Personality the mechanistic concept ceases to be of any practical use.

What is the relation between the earlier (mechanistic) and the later (holistic) elements in composite structures, such as bio-chemical and psycho-physical wholes? How can the material and the immaterial influence or act on each other? This is still one of the great unsolved problems of philosophy, and science finds it no less embarrassing. The tendency for science has as a rule been to look upon the earlier physico-chemical structures as dominant, and upon the later holistic elements of life and mind as essentially unreal or as having only an apparent reality. Science looks upon the physical realm as a closed system dependent only on physical laws, which leave no opening anywhere for the active intervention of non-material entities like life and mind. On this view the activity and causality of life and mind are therefore at bottom essentially illusory. On the other hand, if we have to be guided by our clear and unequivocal experience and consciousness, nothing can be more certain than that our human volition issues in active movements and external actions. Besides, if life and mind were merely ineffective illusions, how could they have arisen and grown in the struggle of existence? While science denies reality to life and mind, the other side retorts by erecting them into vital and mental forces with a substantiality of their own. Thus arises the counter-hypothesis of Vitalism. Both views as a matter of fact are one-sided and misleading; the mechanistic view by ignoring the essentially holistic element in organic or psychical wholes; the vitalistic view by misconceiving the vital or psychic element in such wholes. The fundamental mistake is the severance of essential elements in a whole and their hypostasis into independent interacting entities or substances. Thus body and mind wrongly come to be considered as two separate interacting substances.

In reply to mechanistic Science it can be shown that the holistic factors of life and mind do not interfere with the closed physical system, and that a proper understanding of the laws of Thermodynamics permits of the immanent activity of a factor of Selectiveness and Self-direction, such as life or mind, without any derogation from those laws.

Again, in reply to the Vitalists, who invent Entelechy or some other substantive entity for the system of life and mind, it can be shown that no such *deus ex machina* is necessary; that the fundamental concept of Holism suffices to explain the creative, directive, controlling activity of organic and psychic wholes; and that the attributes of life and mind are inherent in the concept of wholes, and in organisms and humans as wholes. We thus get rid of the notion of separate interacting entities and view organisms and humans as wholes, which involve both the earlier mechanistic and the later holistic phases of Holism. As we have seen Mechanism to be but an earlier, cruder phase of Holism, the problem essentially disappears. A thorough grasp of the concept of wholes and its consistent application to organisms and humans are thus a solvent for the perennial Body-and-Mind problem. We thus envisage the physico-chemical structures of Nature as the beginnings or earlier phases of Holism, and "life" as a more developed phase of the same inner activity. Life is not a new agent, with the mission of interfering with the structures of matter; it involves no disturbance of the prior structures on which it is based. Holism has advanced only one step further; there is a deeper structure, more selectiveness, more direction, more control. But the new is a creative continuation of the old and not a denial of or going back on it. Holism as an active creative process means the movement of the universe towards ever more and deeper wholeness. This is the essential process, and all organic and psychic activities and relations have to be understood as elements and forms of this process. No explanation is possible which ignores this active creative inner whole at the heart of all organic or psychic structures; in the light of this whole all apparent contradictions disappear. This point is further considered in Chapter X.

The fact of Evolution shows that Holism determines the course and the character of the advance. Thus Holism is pulling all the evolving structures faintly but perceptibly in the direction of greater creative synthetic fullness of characters and meanings, in other words, towards more wholeness. The inner trend of the universe, registered in its very constitution, is directed away from the merely mechanical towards the holistic type as its immanent ideal.

How Holism operates in organic Evolution will appear from the next chapter.

AT various points in the preceding chapters I have appeared to contrast Holism with Mechanism and to treat them as opposed processes and points of view. We shall now have to consider their relations more closely, as a proper understanding of these relations will be found to underlie some of the greatest problems both of science and of thought. We shall see that Mechanism and Holism are not necessarily opposed; that both ideas have their proper scope and sphere of usefulness, but that Holism is the more fundamental concept and in its most far-reaching reactions transforms, transcends and absorbs the concept of Mechanism. A proper view of their interactions and inter-relations and of the leading and more fundamental rôle of Holism in comparison with Mechanism is in my opinion important for science no less than for philosophy.

Let me, even at the risk of reiteration, return to what has repeatedly been said before. For me the great problem of knowledge, indeed the great mystery of reality, is just this: how do elements or factors a and b come together, combine and coalesce to form a new unity or entity x different from both of them? To my mind this simple formula of synthesis sums up all the fundamental problems of matter and life and mind. The answer to this question will in some measure supply the key to all or most of our great problems. My answer has already been given; it is in one word Holism. But it is necessary to show how the answer works in detail, and what its relations are to the current and popular view-points which still dominate our science and philosophy. Science and philosophy alike are vast structures, laboriously built up on the basis of certain fundamental concepts. The attempt I am making is to introduce into these elaborate systems a new basic concept, perhaps more fundamental than any of them. And it will be clear that such an attempt must be a most difficult and hazardous one; it involves far-reaching readjustments of settled points of view, the reopening of questions long looked upon as answered and done with, the envisaging of many old problems from a new and novel point of view. To insert the spear-point of the new concept into these vast closed settled systems may at first sight appear a revolutionary, an iconoclastic procedure. But I hope I shall be able to show that this is not really so, that at any rate to begin with the concept of Holism will fit constructively into the work of the past, whatever its ultimate effects may be in the reshaping of these systems on the new basis; that in relation to the old concepts it appears in the field not as an enemy but as a friend and ally in the great battle of knowledge, and that it will help materially in the solution of problems which are practically insoluble on the lines of the old concepts. The concept of Holism is brought forward as a reinforcement at a critical point in the battle, in the hope that it will help to bring victory. But I do not conceal the further hope that in its ulterior effects it will lead to a recasting of much of the situation of knowledge as at present envisaged, and will render obsolete and replace much that is at present considered valuable if not fundamental both in science and philosophy.

How then does the concept of Holism fit into that of Mechanism without directly negativing it, but with the ultimate effect of transforming and transcending it? Let us return to consider our formula once more from this point of view. We have to consider how elements or entities a and b produce the new unity or entity x different from both; and how this involves the concepts of Holism and Mechanism. For the sake of simplicity I take as an illustration for discussion the case where only two elements enter into the new entity, although usually the number of component elements is much larger; the illustration will cover all cases irrespective of the actual number of such elements. I also assume that the concept of Holism has been sufficiently defined and explained in the two preceding chapters to make its relation to the concept of Mechanism clear without further definition. I need only repeat that the concept of Mechanism involves a system or combination of parts in relation to each other, of such a character that these parts do not lose their identity or substantial independence in the combined rôle they play in the system. The system consists of the parts maintained in their identity, and its action is the resultant of the independent activities of all these parts. The parts remain, and the activity of the system is the mathematical summation of their activities. That is in essence the idea of Mechanism—a system or combination whose action can be mathematically calculated from those of its component parts.

Now let us test the application of the concepts of Holism and Mechanism to possible combinations or systems into which the elements or parts a and b enter as components. What are these systems in Nature of which we have knowledge, and how do they exemplify our two concepts? We find the following possible situations:

(1) Elements a and b are material elements in the loosest possible mixture without any active relation to each other; this is the case of a mere mechanical mixture, in which there is no combination of any sort whatever and nothing new arises, and to which neither of our two concepts can be usefully applied. The mixture is arbitrary or mechanical in the vaguest sense, but is not and cannot be called a

mechanism, and it is the negation of the idea of a whole. Mere juxtaposition in space and time is the only description which could be applied to such a situation, which must necessarily be a rare one in Nature.

(2) Elements *a* and *b* are material elements in active physical relation to each other in the combination or system, and this relation affects the characters of the combination. The relation may be one of gravity or electricity or magnetism or any other of the forces by which matter acts on matter. In such a case the resultant system is physical, and may be properly called a mechanism. There is combination of parts, which do not lose their identity, and whose individual actions are summed up and expressed in the action of the system. The ordinary physical categories apply to such a system.

(3) Elements *a* and *b* are material elements which enter into chemical relation to each other, and without losing their identity form a system which is in substance new and different from the component elements. This is a chemical combination of a substantially different character from the physical combination mentioned under (2), which calls for other categories of explanation besides the purely physical ones. As the parts still retain their identity and individual action, the concept of Mechanism applies to their combination; but it is evidently a different kind of mechanism in which a higher degree and intensity of union of the parts are displayed which affect the character and nature of the resultant entity *x*. It is, of course, true that the New Physics is rapidly assimilating chemical categories of explanation to physical categories, but a real difference in the results remains; in character a chemical mechanism is substantially different from a mere physical mechanism, although ultimately the underlying forces of union may be proved to be the same in both cases. Material substances in Nature arise from the combination of both forms of union or synthesis; hence all material substances are properly called physico-chemical mechanisms.

(4) Elements *a* and *b* enter into a combination which

transforms one or both of them so completely that its or their identity is lost and irrecoverable; the resultant entity x cannot be explained as the result of their separate and individual influences and activities; and the merger of elements is far more complete than in the preceding case (3). If this were a complete statement of the facts the concept of Mechanism would not apply here, and it would be a case of Holism pure and simple. But as a matter of fact the energy contents of the elements appear to be at any rate quantitatively reproduced in the new entity x; and besides, x, in so far as it is a material system, still seems to conform to a mechanistic type and arrangement of parts. In both these respects, therefore, the concept of Mechanism may still be partly applied to x. But x is a mechanism of an entirely new type, quite unlike the preceding case (3). It is called a bio-chemical mechanism. But it is a mechanism only in certain respects, and to a limited extent, and of a novel character which necessitates new categories of action and explanation. Beyond that it ceases to be a mechanism and appears to conform to the idea of Holism in all other respects. This is the case where cell a takes in food b, which it transforms into its own system according to a metabolism which differs in material respects from the ordinary mechanical phenomena of physics and chemistry. This is also the case where cell a unites with cell b to form a new entity, in which both a and b disappear finally and irrevocably, and whose character and behaviour cannot be traced mathematically or mechanically to those of a and b. The cases falling under (4) therefore display a mixture of Mechanism and Holism, the relations of which it remains for us to study in this chapter. They form the province of life, and at one end of the vast ladder of life they are much more mechanistic and at the other much more holistic in character. They are the bio-chemical wholes which we shall discuss just now.

(5) The new entity x arising under (4) as a mixed mechanistic-holistic type enters into combination with a new factor of an immaterial psychic character, called Mind; and this, the human type, effects a complete merger of the

biological and psychic elements, with an interaction so close and intimate that the psychic element can only be properly looked upon as an outgrowth or development of the biological characters. In other words, the holistic element which entered into x at stage (4) now becomes inextricably blended with another even more pronouncedly holistic element; and the result is a still further approximation to the full holistic type. In fact man is only mechanistic in respect of his physical bodily organism; the true personality which arises from the blending of the biological and psychic elements into one unique whole is the highest and fullest expression of Holism which Nature has yet realised. If we apply mechanical characters to man's mental or spiritual world, that is only by way of analogy from lower forms of experience, and not because his spiritual structure is in any way of a mechanistic type. Man is based on both worlds; while he has one foot planted on the mechanistic plane, his other is firmly planted on the holistic plane, with a distinct lean-over towards the latter. Essentially he is a spiritual and holistic being, not a mechanistic type, with *sui generis* categories of the mental and ethical orders. But his physiological basis gives him partly a mechanistic character. He is thus what is called a psycho-physical whole. This will be more fully elaborated in its proper connection later on in this work.

This rough summary of the main phases and stages of synthetic development through which inorganic and organic Evolution has passed will suffice to make clear two points which I wish to emphasise:

In the first place, Mechanism as applied to types of evolution is an elastic concept, capable of much refinement in its application to ever higher forms and types. The mechanism envisaged from the point of view of chemistry is different from that of physics, while again the mechanism of the cell and of simple organisms is a vastly different affair from that of chemistry, and even so is stretched to a limit beyond which it ceases in many respects legitimately to apply. We have different levels of Mechanism, with their appropriate concepts and categories of structure and function. When we reach the human stage in its full development in personality, we pass beyond the limits of all possible mechanistic concepts and categories and we find ourselves in the domain of Holism. Mechanism is thus a matter of degree.

In the second place, Holism is also a matter of degree. It begins, as we have seen, as structure; and in its earlier phases as structure it is scarcely different from Mechanism. Indeed we may look upon Mechanism as incipient Holism, as a crude early phase of Holism. In proportion as Holism realises its inwardness more fully and clearly in the development of any structure; in proportion as its inward unity and synthesis replace the separateness and externality of the parts, Mechanism makes way for Holism in the fuller sense. But its realisation is a matter of degree, and there will probably always remain some residuary feature of Mechanism, which will to some small extent justify the resort to mechanistic concepts and categories, even where the most developed and refined Holism is concerned.

It follows from the above that science is not at fault when for heuristic purposes it applies mechanistic methods and concepts to either the inorganic or the biological sciences. Up to a certain point the resort to such methods and concepts is fully justified, and their clearness and narrowing of issues are especially useful for purposes of analysis and research. It is only when the larger holistic considerations behind the mechanisms are ignored, or when mechanistic concepts are pressed too far in their application to essentially holistic structures and functions, that the process becomes harmful and misleading.

It will be noticed that in the synthetic grading-up of the Mechanistic-Holistic process of Evolution, the lower unit always becomes the basis of the next higher unit, becomes as it were the stepping-stone to the next stage. Thus the earlier simpler structure of the atom becomes the unit for the molecule; the molecule for the crystal; the complex of molecules for the cell; the complex of cells for the higher organism; while the still more complex groups of cells become the units for the higher psychic or personal structures. Thus stated the process seems to be merely a regular mechanical, rhythmic series based on the mixture of previous elements. But such a conclusion would be most misleading. The process is not mechanical but essentially creative; at each stage something new arises from the mixture, interaction and fusion of the component elements. But while this newness, this creative novelty arises everywhere, it is at two stages in particular that something utterly new and wholly different in kind and nature arises from the union of the pre-existing elements; those are the stages where so-called life and mind appear; the stages where bio-chemical and psycho-physical wholes make their appearance. These are the two great *saltus* or mutations in Evolution; and it is in connection with them that the great problems of life and mind arise. They are the structures in Nature which do not exemplify pure Mechanism on the one hand or pure Holism on the other; they are double structures apparently exemplifying both at the same time and, what is worse, in a somewhat disharmonious manner. The attempt to harmonise them, to smooth away their discrepancies and to reconcile the contradictory results to which they lead, has taxed the resources of our science and philosophy to the utmost; nor can it be said that the results hitherto attained are in any sense satisfactory. But that should never discourage us from renewing the search for solutions which is given by the very nature of the human spirit.

Now it seems to me that it can be shown that the problems, difficulties and contradictions which arise in connection with these bro-chemical and psycho-physical wholes are due to fundamental misconceptions, and that the application of the category of Holism to living bodies and human personalities will transform the situation and help towards a solution of the apparent contradictions. Let us broadly state the problem as it presents itself from the point of view of physical science and of human psychology respectively.

Now science looks upon the physico-chemical order, upon

physical nature, as it is commonly understood, as a closed system, complete in itself. The chain of physical causation is complete, and there is no need or place for anything of a non-physical character. There is a complete system of equations as between the past and the future. Effect equals cause; and there is no necessity or place for any *tertium quid*. Necessity and determination characterise the order of Nature, the laws of Thermodynamics supply a test of its working character. Where then do life and mind come in? What is their function and their relation to this physical order? What difference can they make to this complete, closed, self-sufficing system? If they have any effect, it can only be by interfering with the inevitable chain of physical causation and thus breaking the laws of energy. If life or will or mind has any practical effect, that would mean an interference with physical causes, with the fixed and determined energy equations. But no such interference can be detected in any direction; the causal physical chain remains unbroken; the laws of energy are unalterable. We are therefore forced to the conclusion that life and mind have no real effect and are of no avail in the world. If they were, the fundamental laws of Nature would be upset. Such is the view-point of physical science.

But, on the other hand, we are just as firmly persuaded by the most clear and unequivocal deliverances of our consciousness that we can choose, that we can direct our attention and action to definite purposes; that our willing is effective; that we can will to perform an act, and perform it accordingly; that our bodily organs respond to the act of will in spite of all the energy equations; that within limits we can do what we will to do. Unless our consciousness and our senses quite deceive us, this seems to be as plain and self-evident as anything in our experience. And thus we are landed in self-contradictions. On the one hand, the unbreakable chain of natural causation and the laws of energy; on the other, our indubitable consciousness of the effectiveness of our power of free self-directed action. How is the contradiction to be overcome? We are not concerned with the hoary old philosophical conundrum of free-will, but the issue is the very live and real one of the fundamental veracity of our clear conscious experience. If in the last resort we cannot believe our consciousness and senses, we had better give up the problem of knowledge altogether.

In this dilemma it seems to me that only one course remains open for us, and that is to accept the direct deliverance of our senses at its face value. If we cannot trust our consciousness when it produces clear, direct and immediate testimony to our power of self-direction and action, how can we rely on it when it proceeds by way of inference to build up a vast construction such as the universal causality of Nature? If we cannot trust our experience where it is perfectly clear and unequivocal, it is useless to attempt to proceed any further in our search for truth. But then the question at once arises, how our minds can act on Nature without breaking the causal chain of Nature. How is the link of Mind inserted into that closed chain? It is unnecessary to discuss at length the answers which philosophers and scientists have attempted in reply to these questions. Science on the whole tends to accept the physical view of natural necessity and to look upon mind as ineffective, as an epiphenomenon which does not avail to alter the course of Nature. It is forced to this view in spite of the difficulty which thereby arises of explaining how this useless and ineffective organ of mind could have arisen in the grim struggle for existence; what biological function it performs and what survival or other value it possesses. But that question need not detain us here. Nor need we consider the theories of psycho-physical parallelism and pre-established harmony and such like, to which philosophers have been driven in their distress in order to explain the apparent miracle of the adjusted co-working of body and mind. There is no doubt that none of these views can be looked upon as satisfactory. And the necessity remains for further exploration. Instead of rummaging in the scrap-heap of philosophy, let us rather explore some new way out amid these historic problems and difficulties of thought. Perhaps our basic categories have been faulty or inadequate; perhaps the facts are all right, and it is only our way of envisaging them, our viewpoints and fundamental concepts, that play us false. The difficulties may be of our own making, and should therefore be of our own un-making and solving. Anybody who has carefully followed our account of Holism in the two preceding chapters will at once appreciate the line of thought which it naturally suggests as the way out of these difficulties. The radical mistake made by both science and popular opinion is the severance of an indivisible whole into two interacting entities or substances, the view of life and mind as separate entities from the body. Life and mind are not new entities which interact with the physico-chemical entities or structures. It is the assumption of these entities and of their interaction with physico-chemical entities of a different order which produces the contradictions for thought and the problems for experience. The assumption of these entities is based on a false view of reality; it leads again to an assumed interaction which does not exist in fact. Between them these two assumptions distort our whole perspective in experience and conjure up for thought a number of contradictions which experience shows not to exist in fact. Thought fails to understand how mind and body interact in a human person; and yet we see the phenomenon before our eyes all the time. Thought fails to understand how the immaterial entity or force of *life* can influence a physico-chemical structure which obeys simply and solely the laws of energy. And yet we see the phenomenon in a living organism all the time before our eyes. It seems inevitable that our experience must be right and our categories of thought must be wrong or inadequate, and that the insoluble puzzles which arise must be due to a misreading of the facts. But, I shall be answered, if life and mind are fictitious and assumed entities, which do not exist in fact, we are back again in the old crude, crass materialism, and the Evolution of which life and mind are the main products and organs becomes a mere hallucination. No, I reply, it is not the reality of life and mind that is denied, but their construction as entities of a character and kind to interact with other entities. It is the false constructions of life and mind and their erection into independent entities which is the source of the trouble and ought to be demolished. A true view of the facts not only will do justice to life and mind, but will remove the problem which a false view has artificially created. And Holism is brought forward as a view-point, a category and an activity which reproduces reality and renders the facts intelligible without distortion or contradiction. Current views of life and mind are wrong; and it is partly to correct these errors that the wide concept of Holism, which includes, underlies and transcends them both, is introduced. Our views of immaterial things have been in process of evolution for thousands of years, and the process is far from complete yet. Remember the view of the soul, held by the Homeric Greeks, as a pale copy of the body; and indeed the present popular notion of ghosts scarcely yet differs from this view. Remember the controversy among the early Christians, of which there is an echo in St. Paul's great chapter in the Epistle to the Corinthians (1 Cor. xv. 35-50)—whether it is the corporeal body or a spiritual body corresponding to it which would be raised to immortality. We still construe life, mind and the soul according to physical analogies or material categories. Let anybody sit down and try to form for himself an idea of the soul's disembodied existence, and he will convince himself how difficult it is to get away from physical analogies, from the pale copies of earthly existences, not very much different from the shades which wander through the cold Homeric Hades. We conceive spiritual things very much on the lines of material things; though conceived as on different planes they are not too far apart, not too different from each other to be able to act on each other and influence each other. It is these conceptions of life and mind as semi-physical entities, reminiscent and redolent of the past, which at bottom underlie many of the great problems of thought. We have really outgrown them; and in a sense they survive as anachronisms and disturbing factors in a world which in most other respects has made the most revolutionary advances in knowledge. They should be reformed and brought into line with the advanced front which is at present held on the battle-field of science.

The vague, popular, ghost-like concept of life is stereotyped and rendered definite by the scientific concept of Vitalism; for our purpose we may take the two as equivalent. Now what is Vitalism? It is nothing but a pale copy of physical force. According to the Vitalists or the Vitalistic hypothesis, a living body is conceived as a material system in which the physico-chemical forces are supplemented by a new force, not of the same character as they, but still sufficiently like them to act on them and to be acted on by them. The Vitalistic hypothesis is right in so far as it considers physico-chemical agencies, considerations and categories as insufficient to explain the phenomena of living bodies. But it is wrong when it proceeds to assume the existence and the interaction with them of a new so-called vital force, which may or may not affect their quantitative relations, which may or may not quantitatively add to or subtract from them, but which somehow has the power to control or otherwise affect the manner in which they are working. A living organism appears to have the power to direct its energies to some definite end, and it will make all sorts of experiments, of trial-and-error co-ordinations of its bodily movements, until it successfully achieves that end. The specific power of directing its energies to certain definite ends or objects or with a certain measure of purpose seems to be characteristic of all living things from the lowest to the highest. This capacity of direction may be conscious or unconscious; it may be reflex or instinctive or deliberate and intentional; but as a phenomenon and a fact of universal observation it is beyond dispute. It is the *explanation* of the phenomenon and the fact which is in dispute as well as its relation to the physical-energy system which it seems to influence or direct. And Vitalism is a theory which attributes this power of inner direction or control to a new sort of force which distinguishes living from non-living bodies. It is, of course, true that with many of the older biologists Vitalism was more a standpoint than a theory; more an attitude of protest against the supposed adequacy and sufficiency of mechanistic or physico-chemical explanation of living bodies, than a definite assumption of a new vital force. They realised that there was something more in the living organism than what could be accounted for on the action of purely physical and chemical forces. In this standpoint they were no doubt right; and in this vague negative sense there is not only no harm, but positive value in the Vitalistic standpoint. But with some of the more recent biologists the Vitalistic standpoint has crystallised into a definite hypothesis which assumes a specific life-force. And it is against this hypothesis that our argument will be directed.

It follows from what has already been said that the very conception of such a "force" is an anachronism, an assimilation of the concept of life to ideas and view-points which are or should be obsolete. It is a question whether the concept of force has any validity at all in physics; whether the dynamical notion of force is more than a mere mathematical notation or terminology with nothing in physical reality behind it. There is a tendency among physicists to discard the idea of force as unnecessary and misleading and to restrict themselves to the concept of energy. Whether they are right or not, it is at any rate clear that the idea of force can only have an application, if it has any at all, in the material physico-chemical order. When it is extended to the province of life, it becomes illegitimate and only serves to materialise what is in its essence non-material and spiritual. The concept of life is already deeply tainted in this and other ways; and that is one of the reasons why I have proposed for purposes of scientific thought and reasoning to discard this vague and abused term, and to substitute instead the notion of Holism, which can at any rate be made clear and definite, and is not vitiated by popular associations and accretions. The Vitalistic hypothesis moves in the opposite direction; by constituting a life-force on the analogy of physico-chemical forces, it tends to materialise life, to hypostatise it into a definite entity, and in this form to set it over against the material body in which it has its seat. Not only is life constituted into an entity interacting with other material entities, but its non-material, spiritual character is reduced to the level of a force among other forces, different from them indeed, but not so different as not to influence them or to be influenced by them. Life as Vitalism or vital force is considered a real entity, and its relations with the rest of the living organism become the source of serious difficulties and contradictions.

I have above briefly stated the naturalistic scheme of science and its sharp opposition to and contradictions of the claims of life and mind as ordinarily understood. That opposition and those contradictions arise from fundamental misconceptions which have their origins in the naïve dualism of our ordinary views of life and mind. Body-and-soul is the model or scheme on which both thought and science are based. There is an *anima* dwelling in a *corpus,* one entity living in close symbiosis with another, and the two profoundly influencing each other. As Descartes formulated it, there is the *res cogitans* in the *res extensa;* there are two distinct separate *res* or entities, and the difficulties and contradictions arise from their mutual assumed interaction. The theory of Vitalism or the vital force seems simply to repeat and to emphasise this dualism. But if we wish to overcome these difficulties and contradictions we have to probe more deeply than these popular views. We must get down to the tap-root from which the two apparent entities or substances must have grown. The subject is most difficult and uncertain; but I may be allowed to put it in the following form.

"Selectiveness," as was pointed out in Chapter III, seems an inherent and fundamental property of matter. Electro-magnetism is a striking instance of that phenomenon; so is the very constitution of matter, whose ultimate forms of structure depend on inherent affinities and selectivities of still smaller structures or units. So is the behaviour of matter in the colloidal state. In the selectiveness of matter we seem to meet with an ultimate property for which no accounting on further more ultimate grounds is possible.

Now selectiveness is likewise the fundamental property of all organism; it is indeed the most primitive property of life. Perhaps it is the very point where the organic and inorganic were still one and began to diverge. A cell shows selective power or selectivity in all its processes, such as the assimilation of its food and the rejection of what is not suitable for its nourishment. An organism shows this selective power in all its movements as well as in its nutrition. There is a selection of ends and an adjustment of its movements to the attainment of those ends. If the adjustment is wrong, if mistaken or abortive movements are made, the experiment is repeated until the object is attained—the food is reached, the danger is avoided, or the enemy is routed. This primitive power of selection or selectivity is not yet choice or will as seen in the higher phases of organic development, but it is the tap-root of choice or will. One form of this selective power is self-direction, which is equally characteristic of organisms. Life has a power of self-direction, of selecting to go in one direction rather than in another, of taking the path which leads to the attainment of its unconscious or consciously realised object. This power of self-direction is clearly only a particular form or species of the more general power of selectivity.

Perhaps I may in passing be allowed to make another venturesome suggestion; and that is, that selectivity is an inherently holistic attribute or quality. A natural whole as a small limited centre of unity has a definite structure which necessarily limits its functioning to certain ways or modes and no others. All possibilities are not open to it; it has only more or less limited degrees of freedom for its activities; it has to confine itself to these and implicitly to reject all others. Anticipating later stages of development, one may say that its choice is limited; in other words, selective action is essential for it. What is perfectly clear at later, more mature phases of Evolution already exists in undeveloped immature form in the most primitive organisms. There is a primitive stage of organic functioning when concepts like will, choice or purpose are clearly not yet applicable, but their root already exists in a sort of organic selectivity or power of self-direction and self-orientation. This primitive organic power of selection is probably not far removed from the inorganic property which I have called by the same name. Apparently the conflict has not yet arisen here.

Let us next consider the most universal generalisations or laws of matter and energy. I refer to the two laws of Thermodynamics, the first of which affirms the universal principle of conservation or constancy of the amount of energy in a closed physical system; while the second affirms the universal principle of the dissipation or degradation of energy. It is these two supreme generalisations which seem to come into irreconcilable conflict with the principles and properties of life and mind, and therefore call for a careful analysis. Now when bodies and souls (including life and mind) are taken as separate entities in interaction with each other, the simplest way of expressing the observed facts is to say that life or mind has a directive power over the body. It was Descartes who first suggested that life or mind might have the power (to use the language of later scientific developments) of directing the energies in a body without affecting their amount, and therefore without a breach of the first law of Thermodynamics. According to this view life or mind in an organism would direct the energies of the body without either creating or destroying any of these energies. We have already seen that this power of self-direction is characteristic of life; and the suggestion was that the exercise of this power, while not interfering with the laws of matter, would explain the influence of life or mind over the body. Leibniz, however, pointed out in answer to Descartes that force (as energy was then called) is not only quantitative but also directional in character. And the second law of motion according to Newton made this perfectly clear. The direction in which a force is acting can be altered only by another force, and this change of direction would therefore involve an expenditure of force or energy. If, therefore, the mind has a directive influence over the body, it can exercise this only by way of adding to or subtracting from the energy of the body considered as a closed system, and would therefore be in conflict with the first law of Thermodynamics. Now in the body as a closed system experiment or observation has never yet shown any such addition or subtraction of energy. The energy put into a living body by way of food, heat or otherwise is always, within the limits of error, equalled by the energy of the work done, the heat produced, and the waste products thrown off. As an energy system the living body is unaffected by life or mind or any other factor of a non-physical character. There can, therefore, be no such direction of the energy of a living organism by life or mind as is assumed; and if there were, the effect would at once be detected in an alteration of the amount of energy in the body. The first law of Thermodynamics, therefore, seems to negative this assumed power of direction of life or mind over the body, and seems to be fatal to any view of directive interaction between the two. Either the first law must be given up, or life and mind are nullities: such are the fatal horns of the dilemma on which we are impaled. But the surrender of the first law is not to be thought of. Although not exactly proved in a rigorous mathematical sense, it is a norm of science which works successfully in practice and which has never been known to be contradicted by any actual observations. It may be that in view of the recent discoveries of the New Physics, which associates the concepts of energy and mass very closely, the law may have to be expanded so as to include both the energy and mass of any closed system.

VII MECHANISM AND HOLISM

But the surrender of the law would bring the whole structure of science toppling down. Nor, on the other hand, is the nullity of life or mind for a moment to be conceded. As I have already pointed out, the sense of effective choice, willing and self-direction is the clearest, most indubitable deliverance of consciousness we have, and its denial must necessarily destroy the very foundations on which experience and knowledge are built. Besides, if Life and Mind are nullities, then the Evolution which produced them must be a farce; but this is totally inadmissible. A way out of this dilemma must therefore be found. But let us first look at the second law of Thermodynamics.

The second law affirms the principle of the universal dissipation or degradation of energy. It likens energy to water; as water constantly tends to run down from a higher to a lower level, so the potential of energy constantly tends to run down, and the energy tends to lose its efficiency and availability. While the energy of a closed system therefore remains constant in amount, it changes in character, it becomes dissipated or degraded and less efficient and useful. And this principle is apparently of universal application in the physical world. When any phenomenon seems to contradict it, that phenomenon will in the end be found to be based on faulty observation.

But living bodies seem to contradict it. In a living body the potentials of energy and efficiency are rising instead of falling. In living bodies complex substances are for ever being built up with a high energy efficiency; and the breaking down of these substances in the processes of life supplies the energy which the living body requires for its proper functioning. These complex chemical compounds with high energy efficiency have been called the high explosives which are necessary for the battle of life. And it is the essential function of living bodies through their subtle metabolism to manufacture these high explosives whose breaking down liberates the energy which life needs for its functions and processes. The process of organic Evolution marks a continuous rise in the complexity of the organic substances produced and the level of the energy potentials reached. Living bodies and Evolution generally, therefore, seem to run counter to the stream of natural tendency as expressed in the second law of Thermodynamics. The systems of life and mind seem to be in contradiction to both the great principles of physical science. Is a reconciliation possible?

Clerk-Maxwell, one of the heroic figures of nineteenth-century physics, was the first to suggest an idea which may open up a possible clue to the solution of the problem. He pointed out that the laws of energy were statistical in character; they regarded bodies, systems and their energies *en masse*, and their principles apply to these energies taken statistically and on an average. When, therefore, the energy of a physical system is spoken of, the average of its particular energies considered together and as a whole is referred to. In this sense, for instance, the principle of the degradation of energy held true, but in no other sense. And he illustrated his meaning by taking as an instance a volume of gas with a certain ascertainable total kinetic energy. In this volume the molecules of the gas would have different energies according to their rates of motion. In accordance with the formula $E = \frac{1}{2}mv^2$, the energy of a particle is proportional to the square of its velocity. Now some molecules would be pushed forward by the impact of other molecules in their line of motion and would therefore have their motion accelerated; others, again, would suffer impacts contrary to their line of motion and would be slowed down. And, as a fact, the molecules constituting the volume of gas would have all sorts and rates of motion and consequent differences of energy. Now, if without introducing any additional energy into such a system, some sifting and sorting out and grading of the different molecules according to their velocities could take place, we could have an assortment of molecules with a higher energy than the average of the gas, while the balance would have an energy below the average. In other words, by sorting out instead of merely averaging we could have bodies with a higher energy potential or efficiency than the average of the mass from which they have been separated or segregated. And this higher energy potential would not be due to the imparting of any additional energy from the outside. Clerk-Maxwell imagined some demon manipulating an aperture inside the volume of gas to effect this sorting and grading, and thus producing a result in apparent conflict with the principle of the second law, which affirms the constant degradation of energy. His point was to make clear that the second law referred merely to a statistical average and was correct only in that limited sense.

But it is obvious that his limitation of the law has a far-reaching significance, and his illustration points the way to the reconciliation of the systems of life and mind with that of physical energy. What if life and mind were conceived as demons of the Maxwell type? We have already seen that their most essential function is selection and self-direction. The sifting, sorting out and grading which Clerk-Maxwell ascribes to his hypothetical demon is the very function of life and mind. Through this selective activity all collision with the second law is avoided, which is true of statistical averages only. And we may go further and show that the self-direction which is inherent in life and mind involves no fresh creation of force or energy in its application to matter, as Leibniz held, and constitutes no infringement of the first law, as is commonly assumed. The same argument which holds for selection (of molecules with a particular speed) in reference to the second law holds also for direction of molecules in reference to the first law. The supposed demon, dealing with our volume of gas would select molecules, not of a certain velocity, but moving in a certain direction, molecules with a certain orientation, in preference to others, and could thus obtain a body moving in a certain direction without the expenditure of any additional energy in bringing about this change of direction. Change of direction need not, therefore, involve any change in the energy situation, as Leibniz held and as is commonly assumed. It is only when bodies are considered as a whole and as averages, and without reference to their detailed structures and arrangements, that the difficulties arise and the physical system seems to come into conflict with the systems of life and mind.

If my reasoning is correct the result is most important. It suggests and indicates the way in which, in bio-chemical and psycho-physical wholes, Life the selector and Mind the director may exercise their essential functions in bodies without coming into conflict with the laws of energy as ordinarily understood. The detailed method and mechanism of interaction is not yet explained, but at least the possibility of conflict is eliminated; we see that these two systems *may* function in harmony and without violation of fundamental physical principles on the one hand or the stultification and nullification of life and mind on the other. The *possibility* of harmonious functioning is established; the *actuality* of the process and its details remains for further discussion.

Let me once more state the issue raised by physical science in connection with life and mind and see how the result we have now reached meets that issue. Taking for granted that the statistical laws of energy apply fully to all purely physical systems, the following questions arise:

1. Do they also apply to systems, such as living organisms or conscious personalities, which are not purely physical systems?

2. Further, in such mixed systems, is the effect of the non-physical factor, life or mind, on the physical part of the system such that the laws of energy do no longer fully apply to this part? In other words do life and mind disturb, deflect and alter the application of the principles of energy to the physical part in such mixed systems as living bodies or conscious personalities?

The answer to the first question is in the affirmative and to the second question in the negative. The laws of energy hold for the physical mechanisms of organisms and persons no less than for purely physical systems; and the influences of life and mind, whatever they may be in other directions, do not invalidate the application of these laws to bodies or persons in so far as they are physical systems or mechanisms.

The laws of life and mind are not in conflict with the laws of energy. An organism is more than a physical structure; but in so far as it is a physical structure it obeys the laws of energy just as if it were nothing but a physical structure.

The result is important, because it does justice to both the physical and the non-physical aspects of bio-chemical and psycho-physical wholes. Ordinarily in the grand tug-of-war between the two aspects in these mixed systems, the palm of victory is awarded to one or the other, according to the naturalistic or spiritualistic leanings of the judges. According to those who adopt the standpoint of physical science, the laws of energy apply to the mixed system, even to the extent of reducing life and mind to the rôle of impotent semblances or mere empty simulacra on the scene of existence. Again, according to the spiritualist view, the factors of life and mind are real and operative, not only on their own proper level and in their own domain, but to the extent of qualifying and modifying[1] even the mechanical relations of the bodily or physical structure, and thus affecting the application of the laws of energy to it. The first conclusion (Naturalism) is contradicted by our direct consciousness, the second (Spiritualism) by the experimental results of observations on living bodies. The reasoning we have followed so far on the suggestion of one of the greatest masters of physical science, has indicated to us how life and mind may discharge their essential functions without impinging on the universal laws of energy, which are the very foundation of the whole system of science. The higher structures of life and mind do not mean the annihilation of the lower structures of energy. Here again, as we have seen before in the general process of creative Evolution, the lower becomes the unit for the next higher; there is a grading of the advance without a destruction of the steps or grades constituting the advance. The higher structure is based on the lower structure without the absorption and disappearance of the latter in the process. Thus mind structures presuppose life structures, and

[1] Hobhouse: *Development and Purpose*, pp. 326, 329.

life structures presuppose energy structures, which are themselves graded according to the various forms of physical and chemical grouping.

The *via media,* the way of reconciliation between the mistaken extremes, which we have followed, is often missed by others because they are misled by hypostatising body and mind as two distinct entities or substances or *res,* as Descartes called them. These two entities or substances are then brought to interact by way of external relations, which are naturally of a mechanical character, as all external relations are. This interaction by way of externality reduces mind to the level of body and thus ends by a practical denial of mind. This mistake is then corrected by the opposite mistake of an attack on the body or the physical order. Thus the Naturalistic and Spiritualistic fallacies arise. Here these mistakes have been avoided by our refusal to look at the two physical and non-physical systems as distinct entities coming into external relations. Clerk-Maxwell's suggestion has taken us right into the inner structure of the gas, and has shown us an inner selective process at work which is by no means merely mechanical, and which has resulted in the segregation of a new structure from the old in a way which constitutes an apparent, but merely an apparent, and no real breach of the universal laws of energy. The fable of the selective demon contains a real truth, and points to the nature of the activity of life and mind in bodily structure. But at best the fable is but a crude and rough version of a matter which requires much more careful exploration. And we therefore proceed now to consider in closer detail the nature of the bio-chemical and psycho-physical unities or wholes, and the relations between the two mixed systems which they include in their wholeness or unity. The best vindication of Holism as a category of explanation would be the light it could throw on the mode of union of the two systems, on the way in which body and life, life and mind constitute unities or wholes such as we know in experience.

Life the selector, and Mind the director, how do they

operate, what is the mechanism which interlocks them and makes them one with the physical? What fundamental conception can we form of the physical, the vital, the psychical which will represent in thought the unities which they are and form in fact? Life starts from the simplest almost purely mechanical forms in the vegetable kingdom and passes upward until it flowers into the marvels of organisation of structure and function, of beauty of form and activity, which we see in the plant and animal kingdoms. And it probably had an immensely long history of development before it attained even the lowest forms now known to us. But all through, the fundamental function of selection, of selective taking and leaving, has distinguished it. Mind again, by selecting the selected, has initiated the power of direction which has gradually evolved into the new world of the free spirit. How can we envisage the physical, the vital, the psychical as together forming unities and wholes as they do in fact?

Naturalism answers this question, as we have seen, by making life and mind the mere unreal accompaniments, the reflexes or shadows, of the real mechanistic physico-chemical system. A solution which in effect rules out half of the world of reality as revealed in our experience cannot be accepted as satisfactory and need not detain us here. Vitalism again puts forward a theory of its own which we may examine for a moment. We shall take it in the form presented by Professor Hans Driesch, who has elaborated a special form of the Vitalist theory with an imposing apparatus of proofs. This is the theory of Entelechy. Driesch supposes a non-mechanical agent at work in psycho-physical systems which has the power of suspending their action in particular respects, thus enabling them to store up and retain their energies, and which again relaxes its suspensory power and thereby allows their energies to be set free and their action to proceed when the situation of life requires it. Where this controlling action on the part of the mysterious Entelechy comes from, Driesch does not profess to know. It evidently corresponds somewhat to

Maxwell's mythical demon. But its power is more closely defined as checking action, when action would mean mere dissipation of energy, and releasing the check when necessary, and thus setting free the stored-up energy of the system to produce the effects, such as we see in the organic world. This relaxing action of Entelechy is non-energetic; it is not the removal of some mechanical obstacle, as such removal would involve some expenditure of energy, however small. The releasing action of Entelechy is entirely an action *sui generis,* just as the suspending action is. Driesch considers that this assumed action of Entelechy is the only possible way in which the causal relation between the mechanical and the non-mechanical world can be made intelligible without sacrificing the fact that organic life is limited by matter.[1] Entelechy is obviously little more than another name for life; life being conceived as a real agent, a real operative factor inside the physico-chemical system which we call the body, and with a real power of action upon it. But as Entelechy is expressly a non-mechanical, non-energetic agent, the mystery of the action of this non-mechanical agent on the mechanical physical body remains entirely unexplained. I fail to see how the concept of Entelechy takes us much further than the fable of Maxwell's demon does. Something like selection, the suspension of action and its relaxation, may probably take place. But the difficulty remains of conceiving how this is brought about and operates. The introduction of the concept of Entelechy does not really help us. We have still to see whether there is anything in the physico-chemical situation which throws any light on the mystery, and whether it is possible to avoid the appearance on the scene of a *deus ex machina,* such as Entelechy undoubtedly is. I shall therefore proceed to inquire what light the concept of Holism, as it has been expounded in previous chapters, throws on this problem of the nature of "life," and of its action on the physico-chemical system which constitutes its body in any living organism. It is unnecessary to point

[1] *Problem of Individuality,* pp. 38-40.

MECHANISM AND HOLISM

out that we are in a region of speculation, where no theories can be brought to the test of decisive experiment or proof. All that we can hope to achieve is to render intelligible what is in itself a great mystery to thought; to supply some possible explanation even if it is not the real one; to suggest a scheme of a possible *modus operandi* which the imagination can visualise to itself. More than a *possible* explanation I do not pretend to give.

Science has made clear, as we have seen in previous chapters, that the physico-chemical system is a structure, a structure composed of elements in more or less of equilibrium. Such is the atom of matter, such the molecule and all chemical compounds which form the substance of living bodies. The equilibrium of the structure is also only approximate; were it complete, little room would be left for change; the physical world would be a stereotyped system of fixed stable forms, and little or no room would be left for those changes and developments which make Nature a great system of events, a great history moving onward through Space–Time. The fundamental structures of Nature are thus in somewhat unstable equilibrium. A change in equilibrium does not mean an alteration in the position and activity of one element of the structure only; there is a redistribution which affects all the elements. It is the very nature of the structure in changing its equilibrium to distribute the change over all its component elements. No demon is at work among these elements to transpose them, to rearrange them, and to vary their functions slightly so as to produce the new balance or equilibrium of the whole. It is an inherent character of the physico-chemical structure as such, and is explicable on purely physical and chemical principles which do not call for the intervention of an extraordinary agent. Another peculiar feature about the change in equilibrium in a physico-chemical structure is that it is never such as to produce a perfect new equilibrium; the new is merely approximate just as the old equilibrium was. We may say that the change is from too little to too much. A structure remains unchanged in spite of a small change in its inner equilibrium; hence the inner instability must pass certain limits before the readjustment in equilibrium takes place. The instance of a supersaturated solution is a case in point, where the solidification or crystallisation lags behind the conditions which bring it about. When the change does come, it again proceeds too far; it swings beyond the necessities of the case; it passes the limits of perfect equilibrium on to the other side, so to say. From too little adjustment it passes to too much adjustment, and again there is a condition of instability which has to be righted by a swing back in due course. Thence arises the rhythmic character of natural change, which links it on to the rhythm of the life-processes, and shows that they spring from the same source in the inner nature of things. Hence probably arise also the definite quantitative increments of change which the New Physics reveals.

This mysterious tendency to equilibrium or inner stability shows the inner holistic character even of physico-chemical structures. There is an internal balance which preserves the type, a push-on when the structure is endangered from one quarter, a pull-up when it is endangered from another. These inner pushes and pulls are not the work of extraneous demons, but represent the inner holistic nature even of natural physical things in their total make-up. And the pushes and pulls are adjusted into a great rhythmic process which becomes the law of life in the next higher grade of structures. We may call the structure a mechanism and its action mechanical. But both ideas are but a superficial view of the real facts, which are so remarkable as to be almost as mysterious as the similar though more complicated phenomena which meet us in the structures of life. Not *laissez-faire*, not utter Chance and Hazard, but control or governance meets us in the inner courts even of physical nature.

I envisage the physico-chemical structures of Nature as the beginnings and earlier phases of Holism, and "life" as a more developed phase of the same inner activity.

Life is not a new agent, with the mission of interfering with the structure of matter; the control which it appears to establish is not a disturbance and upsetting of the natural order. It is itself a structure, based on the lower structures of the physico-chemical order; and the control it introduces is nothing but an extension and development of the natural physical control which, as we have just seen, is already in operation in the lower structures for the maintenance of inner stability. Life is a new structure of the physico-chemical structures of Nature. It is not there to cancel them, to upset or destroy them, but to introduce a still deeper, more fundamental element of structure into Nature. And the structures of the lower order are necessary to it. Without matter no life, without the physico-chemical structures no structure of life. The one is a stepping-stone to the other; nay, more, is an essential element in the other; the physico-chemical structures become the elements in the new complex structure of life. No cancellation, no annihilation, no repudiation of the past; but only more intensive organisation of the pre-existing factors into the new creative structure of life.

The new structure of life differs from the physico-chemical structures which are its material, its constituent elements; the difference is most important and far-reaching, but does not amount to antagonism. A deeper harmony is introduced; the earlier, cruder notes of the physico-chemical order become a new music of being. There is an element of newness, of structural and functional newness, introduced, but the new does not conceal or annul the old. The structural march of Holism has proceeded only one step, one great step forward; but the system and character of its advance remain fundamentally the same. The new is a greater complication, a deeper intensification; there is more selectiveness, more direction, more control; there is more of the whole, of the character of wholeness, in the new structure than in the old. But there is no switching off from the old to the new; the one is a continuation of the other, a continuation indeed of a novel and creative character, but not a denial of and a going back on the other.

Thus life is a structure like matter; and a structure in a similar state of unstable equilibrium. The change of equilibrium has the same rhythmic character; only this character is far more noticeable and pervasive than the similar phenomenon in matter. The rhythmic oscillation becomes the distinguishing mark of the functions of the life-structures. The pulsations, the rhythmic flow of the functions of cells form the law of life, and incidentally become the basis of the new element of music in life; they give to music that primordial fundamental character which takes us back to the very beginnings of life on this globe, and makes music the deep appeal of all the long ages to emotions the most primitive as well as the most highly evolved. The rhythm of equilibrium shows the close linkage between the physical structures and the life-structures. And its music links all life together through all the ages.

The equilibrium of the life-structure also gives us the origin of the idea of life as the selector, the suspensor and relaxor of the activities of the new structures. In any change in the equilibrium of the physico-chemical structure there is, as we have seen, the distribution of the change over all the component elements; there is the new arrangement and alignment of elements and their activities which conduce most effectively to the balance of the whole. This is exactly what happens in the rhythmic change of the life-equilibrium. In the movements of that change, elements are rearranged, functions are readjusted with a view to the conservation and activity of the whole. Selection and direction and control are inherent activities. No extraneous factor does this; no mysterious stranger needs to be introduced from some alien world to work the mechanisms. It is the very nature of the equilibrium of the new structure thus to direct and regulate—thus to transpose and distribute the factors of equilibrium among its component elements, thus to rearrange and readjust and interchange elements of structure and function so as to constitute its new balance of structure and function, and to preserve it as a whole. The selective regulative nature, character and activity of life arise from the very nature and process of the equilibrium in the new structures which we associate with life. The conception of Entelechy is therefore not necessary. The regulative equilibrium of the new structures which we call organisms is sufficient. This equilibrium oscillates between certain limits, and within these limits the structure maintains its balance of parts and activities inside the physical system of Nature. Beyond those limits it is, of course, destroyed, and the structure of life is therefore most closely and intimately associated with the conditions and properties of its material medium. It is not an independent entity, self-created and free from the trammels of matter. It is a complex structure of the simpler structures of matter, and therefore dependent on those structures and their laws. But within certain limits it creates internally its own adjustments as a structure and is to that extent free from matter. It is more of a whole, it has a measure of freedom, and in its self-maintenance and dynamic stability it shows a power of internal regulation and co-ordination which is quite beyond the range of the lower physical structures. Take, for instance, the manner in which the bodily temperature is maintained under all sorts of conditions through a most minute and delicate co-operation of a vast number of physiological factors and mechanisms. Professor Haldane has very ably dealt with this aspect of the matter and has shown with great force that Physiology demands imperatively new categories of explanation, and can no longer rest content with the crude conceptions of mechanism which have so far been prevalent. But, on the other hand, his argument must not blind us to the fundamental similarities between inorganic and organic structures. Organic structures do but repeat on a higher plane of organisation and with an added element of newness, inherent in Holism, that process of self-adjusted, self-regulated equilibrium and of inner self-control which equally characterises inorganic structures.

I have to make one more assumption in regard to the character of the new structures of life and their change of equilibrium. Evolution is a fact of observation and experience; and it shows a persistent trend. From matter to life, from life to more life and to higher life; from higher life to mind, from mind to more and higher mind, and to spirit in its highest creative manifestations. There is a process with a persistent trend, which cannot possibly be the mere result of accident. If it were all a matter of chance and contingency, the odds would be infinitely more in favour of chaos than of this persistent trend of intensifying structure and order. The creative process of Holism consists in the intensification of structures, in small elements of newness appearing in existing structures until the basis is thereby laid for a new departure in structure; but still on the basis of the pre-existing structures, and so to say in line with the pre-existing structures. Matter and energy were probably such departures in structure from pre-existing structures which have now passed away and are unknown to us. Similarly life is a new departure in structure, but still in harmony and more or less in line with matter, whose laws are only in appearance and not in reality opposed to the processes of life. Life, again, represents a rising scale of structure until the foundations have been prepared for a new departure in structure in the form of Psychism or mind; and mind is on the whole in harmony with life and in line with it. There is thus a persistent trend in the evolution of structures and of the forms and types of existence: how is this to be accounted for on our theory of the dynamic self-controlled equilibrium of structures?

I can answer this question only by the assumption of a persistent overbalance of the equilibria of existence in a more or less definite inner direction or with a more or less definite orientation. In other words, the rhythm of change in equilibrium is not perfectly reciprocal, but the oscillation always tends to be somewhat more in one direction than the return swing in the other direction. Thus there is an infinitesimal overbalance or tendency to increased overbalance in the one direction. And this is the direction in which the dynamic equilibrium is therefore continuously moving, and which determines the persistent trend of Evolution. Why should there be this difference? Why are the oscillations of equilibrium not equal and opposite? I can only make the further assumption, in keeping with the spirit of our whole subject, that all structures are under the fundamental influence of Holism, which is faintly but perceptibly pulling all structures in its direction. The trend of this slight overbalance of equilibrium is thus towards Holism, towards a structural character which will ever more approximate towards wholeness. In other words, the inner trend of the universe, registered in its very constitution, is directed away from the merely mechanical towards the holistic character and towards the realization of Holism as its immanent ideal. The nature of the universe points to something deeper, to something beyond itself. The persistent overbalance in its equilibrium shows that it is not self-sufficing. It has a trend; it has a list. It has an immanent Telos. It belongs to or is making for some greater whole. And the pull of this greater whole is enregistered in its inmost structures. I only mention this subject in passing, without waiting to develop it more fully at this stage.

At the conclusion of my argument I shall be asked how the result bears on the problem of Mechanism and Holism with which I began this chapter. Life has been shown to be a structure, or structure-like, or best represented by the imagery of a structure, just as matter is. Life has appeared as a continuation on a higher plane of the sort of structure which matter is on a lower plane—a higher structure of the same material, and therefore at bottom not something utterly alien to and different from it. And I shall be asked, "Is Mechanism then final? Is life only a more refined mechanism, a mechanism of a higher type, but still a mechanism? And is Mind a still more refined mechanism? If not, then where in the progress of my argument does the Mechanism come to an end and the Holism begin? Where is the great break, the great rift between the material and the non-material which experience reveals? Or is experience at fault in accentuating this great break or rift?" In answering these questions I shall not go back to the preceding argument, but I shall briefly state my general impression, my standpoint in this matter, which is both the source and the outcome of the preceding argument. Mechanism is not final. It is not all Mechanism at the beginning, nor is it all Holism at the end. If the two have to be distinguished we may say that they vary in inverse proportions with the forward march of Evolution. But the deeper view does not distinguish them, and discloses the fundamental unity. Mechanism, as I have said, is a phase, an earlier immature phase of Holism; just as life is an intermediate phase, and mind is a later phase; while other phases are probably in store for the experience of the higher race which will succeed the human in the future. Holism is a mediating concept; it is the reality which underlies all the phases. And in the self-fulfilment of Holism, one phase passes into another. And the past phases endure, though in ever diminishing degree, in the newer phases. Hence I have adopted the imagery of the structures, ever more complex structures embracing the same material as the earlier structures, together with these earlier structures which are the units of the later structures. I have adopted this structural imagery because thought is relational or structural, and therefore more easily grasps elements of structure than mere abstract principles or tendencies; and because in its interpretation of the physical world thought has already adopted the imagery of structures. I have therefore represented Holism as structural at all its known phases, and have distinguished the phases as differences of, and advances in, structure. But structure does not mean Mechanism. Mechanism is but one form of structure.

The structure of mind is not mechanistic, nor is that of life. The fact is that there is an insensible passage of change from the earlier to the later, but that the change is never complete and that something of the earlier mechanistic phases survives in the later spiritual phases, which are essentially non-mechanistic. The passage is a creative one at all stages, elements of the new are continually appearing, but on the whole so minute as to escape notice. It is only at certain stages that the new appears to be not only sensible but striking. And here our experience seems to have magnified the change by hypostatising the new into a distinct substance or entity, and placing it in opposition to the old according to a fundamental polar tendency or polarity in all thought and experience. In measuring and reading-off reality we must make allowance for the small eccentricities of our instrument of thought, and we need not on that account discredit the instrument itself. If we make the small allowances necessary within the margin of essential error, we find the breaks and gaps and hypostatised distinctions are smoothed out and accounted for, and there remains one great fundamental Process creatively flowing forward and giving to all the manifold and diversified forms of existence the unity which is theirs by inalienable birthright. That Process is not a mere ideal; it is already, in some measure, a fact, accounting for all the particular facts and things of the actual universe. It is Holism, and it should not be confused with the many other things which are its legitimate or (where thought errs) its illegitimate offspring.

CHAPTER VIII

DARWINISM AND HOLISM

Summary.—Darwin's conception of Organic Descent and his formulation of its laws were the beginning of one of the most far-reaching revolutions in human thought. Holism gives a new view of one of the Darwinian factors, and extends the scope of Evolution beyond the purely organic domain.

Darwin traced Organic Descent to the interwoven effects of two factors: an inner creative factor, Variation, operating spontaneously and somewhat mysteriously inside organisms and modifying their hereditary structures and functions in very slight degrees; and an external factor, Natural Selection, which operates selectively on these slight variations, weeding out those organisms whose variations were less suitable to their environment, and leaving the organisms with suitable variations to multiply and develop. By continuous summation of small useful variations through many generations definite specific characters would in time be achieved and new species arise.

Darwin laid most stress on the factor of Natural Selection; on Variation he was vague and hesitating, but there is little doubt that he included not only inborn variations but individually acquired modifications among the elements which ultimately become specific characters. Thus all the multitudinous forms of life would in the end be moulded by both factors into very close conformity and adaptation to their conditions of life.

The great Darwinian conception has been somewhat blurred by later developments, in which attention has been concentrated on the factor of Variation rather than on Natural Selection. First Weismann denied the transmissibility of acquired characters, and thus made it difficult to understand how organisms through their experience and habits of life become gradually fitted and adapted to their environment. Then De Vries eliminated all small variations from the account and attributed all specific advance to large well-marked "mutations" occurring very occasionally. This made it still more difficult to understand slow age-long adaptation, for instance, to habitats and ecological conditions. Finally, the Mendelians or Geneticists have developed the conception that in organisms there are well-marked stable unit-characters whose combinations in crossing follow a certain definite law; and the experimental Evolution and Cytology of to-day consist mostly in tracing these unit-characters and their manipulations in breeding and in the laboratory. The idea of more or less mechanical combinations thus takes the place of the idea of creative variations, which underlay the Darwinian conception, and it becomes most difficult to understand how the new variation arises, and how it is that Evolution is really progressive and creative, and not a more or less stationary régime of casual character combinations.

These later developments take too narrow a view of Evolution as a whole and therefore tend to become one-sided and to over-emphasise certain aspects of the whole process. They are, however, right in their emphasis on the inner creative factor which is the real positive motive force of Evolution. The real secret is in the cell, in the germ-cell or fertilised ovum rather than in the external situation, important as that is. That is the inner seat of Holism, which is the real source of all variation and Evolution.

There is, however, no doubt that variation is influenced directly by external ecological conditions, which show themselves in the general characters of plant formations and societies, for instance. And there can likewise be little doubt that acquired characters in the long run reach down to the hereditary germ-cell and become transmissible variations. While these variations are still small and without survival value the acquired characters and animal routine shield and nurse them until they are strong and developed enough to confer survival value on their organisms. Modifications thus are the rough material of variations; and to that extent Weismann was wrong, and Darwin—and further back, even Lamarck—right.

There is, however, a further complication which cannot be dealt with on purely Darwinian principles. Modifications and variations do not come singly but in complexes, involving many minor and consequential modifications and variations. Are they all individually "selected" even before they have any survival value or strength? These difficulties force us to look deeper, to abandon the idea of the individual selection of variations, and to look upon the advance as not being that of a single variation or variations but of the organism as a whole. It is the organism that advances on a certain more or less limited front; the "variation" is only the most conspicuous point of advance, but there is a whole curve of advance involving many other minor points. In other words, the advance is holistic and the variation is only the most striking item of a whole series. And the progress and survival of the variation are an equally holistic affair. The organism is simply maintaining its own advance in the variation; the variation issues from it and is in conformity with its whole trend and movement; the variation is not single and unsupported, but behind it is the whole force of the organism of whose inner movement the variation is but the most tangible expression. It is thus the organism as a whole which in the first instance "selects" the winning variation or series, and confers on it support and survival value. "Holistic Selection" is therefore in operation at the birth and through the early nursing stage of the variation, and it is only at its maturity that Natural Selection takes over, and the variation begins to fend for itself, so to say.

Holism must likewise be called in to explain organic co-ordination. It is, for instance, impossible without it satisfactorily to explain all the innumerable co-ordinations and co-adaptations in structure and functions which constitute the action of a living organism. No merely mechanical explanation of co-ordinated animal movements or action has even been given. The animal acts as a whole, with a unity and effectiveness of action which is no mere mechanical composition of its movements. Holism not merely as a concept, but as a real factor, is necessary to account for this unique unity of organic or psychic action.

Holism is not merely creative of variations, but just as much repressive of variations. It is as often inhibitive as creative; it holds in check certain features while it releases and pushes forward others. Thus the balanced whole of the Type is achieved. This repressive aspect of progress is neglected by Darwinism, but it is just as real as the active variation. Both together underlie the types and structures of life. This repressive tendency, already fully at work on the organic level, becomes much more conspicuous on the psychical level, where it operates as ethical restraint, so essential in the formation of the Personality as a moral whole.

From the holistic point of view it can be shown that the inner and outer factors in Evolution lie much closer together than is commonly thought, and the grandeur of the Darwinian vision, instead of being dimmed, stands out in even greater fullness.

Finally, Beauty in Nature is holistic, is of the whole, comes from Holism, and is explicable on no other principle. Holism thus accounts not only for the origin of forms and types, but also for their Values, which far transcend the survival values necessary merely for the utilitarian purposes of Nature.

NEWTON's Law of Gravitation is perhaps the most striking instance in the whole history of science of one simple generalisation bringing within its sweep the widest array of physical facts. The new heliocentric point of view had already become generally accepted when this law was formulated, but vast masses of facts remained which could not be co-ordinated, and required explanation from the new point of view. The Law of the Inverse Square, as laid down by Newton, was completely effective. The phenomena of falling bodies on this earth, the motions of all terrestrial bodies, the movements of the solar system and of the starry universe as a whole, many of the phenomena of physics as known and understood at that time—all seemed to find their correct place and explanation under this all-embracing formula.

Newton did not pretend to understand or explain gravitation itself, and his lifelong meditations on this profound problem afforded him no clue as to the nature of gravitation. But the law of its action, the phenomena which happen on its assumption, he formulated with a simplicity and effectiveness which made it another instance of Columbus' egg. In the way of all human matters the law itself came to be looked upon as more than a law, as an explanation, indeed as an operative factor explaining all the phenomena which it covers. And it is only in our own day that gravitation in this sense has been shattered, and its law as formulated by Newton has come to have a restricted application. Relativity has dethroned gravitation, and for the moment Einstein's Ten Equations rule the universe, where before the equation of the inverse square was the only and unquestioned code.

Immanuel Kant, himself one of the great kings and legislators of thought, looked upon the Newtonian system as final; he raised the vision of some future Newton who would discover and formulate the laws of life, as Newton had laid down the laws of motion and of matter. Beyond all doubt Darwin fulfilled that vision, not perhaps in the sense intended by Kant, yet in a way which has made him perhaps an even more epoch-making figure than Newton. Newton proved epoch-making for science, while Darwin has become epoch-making in a far more fundamental sense. He has changed our whole human orientation of knowledge and belief, he has given a new direction to our outlook, our efforts and aspirations, and has probably meant a greater difference for human thought and action than any other single thinker. But even he is not final. He even less than Newton is final. He has pointed the great way, and on that way we are already travelling beyond his great vision.

Let me first state Darwin's law, which was just as simple as Newton's, and much more easily intelligible. Among living beings there is a tendency to vary and over-multiply; in consequence, a struggle for survival becomes inevitable, and in this struggle for existence the fittest survive. This explains the origin of species and all organic differences in the world. The tendency to variations is a fact patent to everyone; so is the over-multiplication of individuals under favourable conditions and in the absence of external restraints; the resulting struggle which Darwin calls Natural Selection is well known to everyone who has the least knowledge of animate Nature. These are the simple bricks of fact with which the Darwinian theory is constructed. Surely as striking a case of Columbus' egg as was ever presented. The genius of the Master was shown by the vastness of the structure he produced from these simple materials of common-sense and common experience. From these simple commonplace facts he explained the infinite variety of the forms of life which occupy the earth, their geographical distribution both in the past and in the present over the face of the globe, and the marvellous closeness of their adaptation to the physical and other conditions among which they live—adaptation to land and sea, to fresh and to salt water, to conditions of soil and climate embracing the extremes of heat and cold, to the widest range of wet, arid and desert conditions, and to all the innumerable facts and situations which lead to the interweaving of the mysterious web of life.

The whole Darwinian theory is summarised in the last sentences of the *Origin of Species* with a simplicity and beauty of statement worthy of the simple but profound genius of the Master, and they raise before us in a few touches the great Darwinian vision. They have often been quoted, but will bear re-quotation here, and for all time:

"It is interesting to contemplate a tangled bank, clothed with many plants of many kinds, with birds singing on the bushes, with various insects flitting about, and with worms crawling through the damp earth, and to reflect that these elaborately constructed forms, so different from each other, and dependent upon each other in so complex a manner, have all been produced by laws acting around us. These laws, taken in the largest sense, being Growth with Reproduction; Inheritance which is almost implied by reproduction; Variability from the indirect and direct action of the conditions of life, and from use and disuse; a Ratio of Increase so high as to lead to a struggle for life, and as a consequence to Natural Selection, entailing Divergence of Character and the extinction of less-improved forms. Thus, from the war of Nature, from famine and death, the most exalted object which we are capable of conceiving, namely, the production of the higher animals, directly follows. There is grandeur in this view of life, with its several powers, having been originally breathed by the Creator into a few forms or into one; and that, whilst this planet has gone cycling on according to the fixed law of gravity, from so simple a beginning endless forms most beautiful and most wonderful have been and are being evolved."

I am free to confess that there are few passages in the great literature of the world which affect me more deeply than these concluding words of Darwin's great book. They have a force and a beauty out of all proportion to their simple unadorned phrasing. They are the expression of a great selfless soul, who sought truth utterly and fearlessly, and was in the end vouchsafed a vision of the truth which perhaps has never been surpassed in its fullness and grandeur.

Darwin assumed two operative factors in the organic world: (1) Variation in the reproduction and inheritance of living beings, and (2) Natural Selection, or the survival of the fittest, as Herbert Spencer called it. Darwin's name is principally associated with the second factor, with which his works mostly deal, and which he elaborated with an unrivalled wealth of detail. He devoted much less attention to Variation, and indeed used it chiefly as a peg on which to hang his theory of the origin of species through Natural Selection. Variation was to him a mysterious fact for Natural Selection to work on. But its spontaneous uncontrolled character puzzled him. He found no helpful imagery to explain the puzzle. He suggested the theory of Pangenesis, which showed great insight, but it has not been adopted by his successors. The germ-cell theory, which supplied a mechanism for heredity and variation alike, was a later discovery. The science of Genetics has mainly arisen since his day. Not only were his views on variation meagre and vague, but such views as he had have not been adopted by later Darwinians. Thus in the passage just quoted he attributes variation to the "direct and indirect action of the conditions of life," and to "use and disuse." Most Darwinians to-day hold very pronounced views in the opposite direction, and deny that these are the sources of Variation. At present there seem to be indications of a reaction, of a return to Darwin and even to Lamarck, and a tendency to look more favourably upon the views of Darwin on this important point. But the fact is that Darwin is on the whole vague on the subject of Variation, and concentrated all his strength on the other principle of Natural Selection and its effects in shaping the organic world.

However this may be, there is no doubt that both Variation and Natural Selection are essential elements in the Darwinian theory. Darwinism, in fact, implies two factors: an internal factor, operating mysteriously in the inmost nature and constitution of living organisms, and an external factor working along independent lines on the results achieved by the internal factor. The inner factor, Variation, is positive and creative, producing all the variations which are the raw material for progress. The external factor, Natural Selection, is essentially negative and destructive, eliminating the harmful or less fit or useful variations, and leaving the more fit or useful variations free play to continue and multiply, and in this process fitting and adapting the individual to the character of its environment. As De Vries has phrased it, the inner factor explains the arrival, and the external factor the survival, of the fit or useful variation or organism.

Darwin's over-emphasis of the second or external factor had one very unfortunate result: it directly and powerfully reinforced and exaggerated the mechanistic conception of the universe. The *vera causa* of organic change and progress appeared to be Natural Selection, an external factor operating on organisms *ab extra*, in the same way as physical or dynamical forces are impressed on bodies or their parts from the outside. Mechanical analogies began to be applied, and Evolution came to be looked upon as the mechanics of organic development—*Entwicklungsmechanik*, as it has been called by Wilhelm Roux. The whole tendency of Darwinism has therefore been vastly to add to the dominance of the mechanistic hypothesis, which has through it come to extend its sway from the kingdom of matter to that of life. What is more, the simplicity of the Darwinian theory has helped to make, not only Evolution, but the mechanical view of Evolution, common property. The mystery of progress seemed to become quite simple and intelligible on this theory. It all depended on the survival of the fittest, and the survival of the fittest was so simple and clear an idea, and one too so deeply rooted in our ordinary empirical experience, that it seemed all a matter of course which had only to be pointed out by Darwin to be accepted by everybody. The difficult part of the theory, the aspect of it which even to Darwin had remained a mystery, the inner creative factor of Variation, was ignored while Darwinism was in the course of being generally accepted, and accepted in the mechanical sense.

This was the first phase of Darwinism, the phase during which Natural Selection was chiefly stressed and was the dominant note in the theory. Then came the second phase, when attention began to be given to the other factor of Variation. With this Neo-Darwinian phase the name of Weismann is for ever honourably associated. Many great labourers there have been in this field, but the name of Weismann will ever stand out pre-eminent as the biologist who, whatever his mistakes in detail, initiated and developed the exploration of the germ-cell as the source of Variation in Evolution. Weismann turned the gaze of Evolutionists from the outside to the inside of the process, from the apparent mechanism of external interaction and clash to the mystery of the inner process. And what he taught was not only most surprising, but remains one of the most significant and important truths in the whole range of biology. I shall deal with this matter just now. But before doing so I wish to point out that Weismann and his fellow-workers were handicapped in their labours by the mechanical view of Evolution which had already become a fixed dogma in the earlier stage of Darwinism. If anywhere, the mechanistic conception should have received its quietus in the domain of Variation, in the exploration of the inner process or factor of Evolution. Unfortunately Weismann and several of the most prominent biologists who developed this second phase of Darwinism arrived at their task not only as convinced Darwinians, but as mechanistic Darwinians.

The great battle in which Darwinism had won was tacitly considered a victory for the mechanical view of it. And thus the whole problem of Variation, as viewed by these leading Neo-Darwinians, came to be one of investigating or finding the *mechanism* of Variation. Their services have been great, and the route they have opened up will in the years to come lead to even greater results. But there is no doubt that the mechanistic conception has been a grave handicap to them, and that many of their errors are directly traceable to its disturbing and distorting influence. In the first chapter I tried to show how erroneous the conception of Natural Selection as a purely mechanical factor in Evolution was. In this chapter I shall endeavour to show that the purely mechanistic conception of Variation is just as arbitrary and misleading.

The root of Weismann's difficulties lies in his mechanistic conception of the germ-cell. The cell, as we saw in Chapter IV, in its metabolism already shows many of the functions and activities which we associate with the complete individual organism. It is itself a holistic individual, with the most marvellous selective and regulative powers, reminding us (on a much lower plane) of what at a later stage of Evolution appears as the psychical factor. This applies to the germ-cell even more than to the ordinary body-cells. The germ-cell has its "field," and the field of the germ-cell is much more important than is ordinarily thought. Experimental Evolutionists seek more in the actual structure of the germ-cell than is there. There is much more in the inheritance of the germ-cell than can be identified by an analysis of its structure. And this *more* is in its field, which represents that part of the germ-cell which has not yet been crystallised and hardened into sensible structure. Much of its past and its future is in its field; in its field the creative adjustments are begun which are ultimately translated and incorporated into its structure. Here as elsewhere the field is the area of becoming, of creativeness, the growing surface of the structure. To confine our view of the germ-cell to its apparent structure is simply to atomise our conceptions on chemical analogies, and to narrow them unduly to the neglect of very important features in the functions and activities of the germ-cell, and to compel us in the end to adopt that mechanical view which is the negation of its inmost nature as a living holistic individual unity. In this chapter I shall endeavour to show how the concept of Holism acts as a solvent for the difficulties created by the mechanistic conception of Variation.

The alterations in the Darwinian scheme introduced by Darwin's successors have had a profound effect on that scheme as a whole; so much so that it is to-day difficult to say how much of Darwin's great vision still survives. In order to realise this, it would be advisable to compare Darwin's general ideas of the facts of Variation with the modifications introduced by his successors.

In Darwin's view, it was not only the operation of Natural Selection that was moulding living things in conformity with their environment, by eliminating those that were less suited to the conditions of the environment. Variation was also bearing its share in this process of assimilating and adapting them to the environment. The close fitting of species to their habitats and environmental conditions which is so distinctive of animate Nature was, according to him, the combined effect both of Natural Selection and Variation.

In order to ensure clearness in what follows we have to distinguish between various forms of so-called "variation" in living things. In the first place, we have *modifications*, which are due to the functional activities and experiences of the individual in its own life, and not to inheritance from parents or ancestors. The effects on the bodily organism or on particular organs of their use or disuse in any definite way would be such modifications. An animal changes its mode of life and in consequence ceases to use certain organs, or begins to use them in a new way or for a new purpose. Such disuse tends to the atrophy of these organs, just as such new or increased use would develop them. Such atrophy or development respectively in the bodily organism is a modification. All changes or characteristics acquired during the individual life are modifications.

In the second place we have *variations*, which are small changes passing by inheritance, and not due to the developments or acquisitions of the individual life. A small alteration from the type which an animal has inherited from its parents is a variation, in contradistinction to a modification which has been brought about in its own lifetime. In the third place, a large very marked inherited change is called a *mutation*. Any inherited change large and marked enough to constitute a new variety or species is a mutation.

Now I think it is beyond question that according to Darwin's view all three forms of change—modifications, variations and mutations—were useful and operative in the ultimate production of new species. Modifications due to individual use or disuse he certainly pressed into the service of his scheme of Evolution; and although it is not quite clear how far other modifications were similarly treated by him, it follows from the above quotation as well as from other passages in his works that variations due "to the indirect or direct action of the conditions of life," in other words, alterations affecting the individual life, could, to an extent never clearly defined by him, avail for the production of new species. As regards mutations, while he gave reasons for disbelieving in great and sudden changes as the ordinary rule of evolution, it can certainly not be said that he excluded them. His view was that the slow and gradual summation of small modifications and variations continuously conserved or kept going by Natural Selection would, and in fact did, in the course of many generations amount to a sufficiently large and marked change to constitute a new type or species. The continuous summation of the effects of use and disuse and the other conditions of life, as well as the accidental inherited variations which were of a more mysterious origin, would necessarily co-operate with Natural Selection in bringing about the close adaptation of the species to its environment. The result was the vast and intricate system of adaptations and co-adaptations, of harmonious adjustment between Nature and organic life, ramifying through the infinite details of the web of life which we see in Nature. Thus Evolution was explained, thus all the fine adjustments and adaptations in Nature were explained. Only a very long time was required for the infinitesimal calculus of Natural Selection to produce the various results, and that requirement was conceded by the astronomers and geologists. Darwin's view seemed very well to fit in with the fossil record as well as with the facts of geographical distribution, which he looked upon as the keystone to the laws of life. No wonder that the appeal of Darwin's theory proved irresistible and its effect crushing on all the older points of view. The triumph of Darwin's splendid vision of Evolution seemed complete.

Then the second phase of Darwinism began, with the detailed search for the methods and mechanism of Variation and with the venue shifted from the ample range of Nature to the research laboratory of Genetics. First Weismann negatived the inheritance of acquired characters, and of modifications due to use or disuse or other environmental conditions operating on the individual life. Only the accidental germinal variations, and none of the moulding effects of the environment on the individual, could avail in the building up of the new species. Then De Vries came forward and largely eliminated small ordinary variations from the account, and thus practically confined progress to mutations. Finally, the experimental Mendelians or Geneticists appeared, and through their researches and experiments appeared to confine Evolution to the interchange, the combinations and permutations of definite existing unit characters. The combined effect of these three advances on the Darwinian theory might appear largely destructive of Darwinism itself. If, following the Mendelians, we hold that the interchange of definite pre-existing unit characters is all there is in the process of Evolution, advance becomes impossible and creative Evolution disappears. If, according to De Vries, accidental mutation is in a large measure all there is for Natural Selection to work on, the advance becomes indeed a most precarious affair, instead of that steady, continuous, delicate process which has been going on through the geological ages. If, according to Weismann, modifications from use and disuse and similar causes have no survival value and are inoperative in the formations of new species, it becomes most difficult to understand the universal close-fitting adaptations of species to their conditions of life. For there is nothing in common between the accidental variations and Natural Selection, and there is no clear reason why or how this clash should not produce chaos and disaster, rather than the harmonies and adjustments which actually characterise the relations of animate and inanimate Nature. Darwin's theory, even if it were wrong in its details, certainly served to explain and render intelligible the broad facts of the order, adjustment and progress observable in animate Nature. His successors' theories, even where they are correct in detail, fail to explain these facts, and make of the world of life as a whole an unintelligible and in some respects an incredible affair.

It would, however, be a serious mistake to look upon the more recent developments in the nascent science of Genetics as covering the whole wide field of the Darwinian theory. So far as I know, they have no such scope, nor are they so intended or understood by those who are responsible for the very important researches in Genetics now being successfully carried on in biological laboratories. These researches are intended to follow up a special line which was first opened up by the experiments of the Abbot Mendel of Brünn in the time of Darwin. They occupy a very restricted area of the whole field of organic Evolution, and are really concerned only with the elucidation of the special set of problems arising from the crossing or hybridising of races, varieties or definitely distinct variations. Those problems centre around the important question how biological characters *already in existence,* whether patent or masked, behave when brought into contact with each other. Mendel found that certain existing characters behaved as firm and stable units, very much as atoms or molecules do in chemical combination, and he also discovered the law of the proportions in which these unit characters are reproduced in the offspring. Thus if individuals of dominant character a are crossed with individuals having recessive character b, then in the second filial generation the numbers of individuals respectively with a characters, and b characters, and mixed a and b characters are given by the algebraic formula

$(a + b)^2 = a^2 + 2ab + b^2$. In other words, 25 per cent. of the second generation will be pure a's and pure b's respectively, and 50 per cent. will represent individuals with mixed qualities, which on being again crossed with each other will again produce pure a's and b's and mixed ab's in the same algebraic proportions; and so on apparently *ad infinitum*. His researches have been amply confirmed by later inquiries, and they have also established that not only do these unit characters behave as fixed and stable entities, but, very much in the manner of radicle groups in Chemistry, groups of such unit characters also sometimes behave as stable combinations, and enter into combination with other unit characters as persistent unities. This is all very remarkable and interesting, and has important bearings on the practical improvement of breeds and races of animals, and on the beginnings of the new science of Eugenics. But for our present purpose it is merely necessary for me to point out that Mendelism or Genetics deals with the manipulation of existing characters, and not with their origin, genesis or creation. The main question before organic Evolution, how specific characters are produced which have not existed before, is not directly touched by Mendelism. The problem of the creativeness of Evolution in the origin of species, and in organic progress generally, lies beyond the province of Mendelism. Mendelism deals with results already achieved by Evolution, and not with the creative process by which they are achieved. No doubt it may and in due course will incidentally throw important sidelights on the mysterious creative process; but it will probably be no more than sidelights. In other words, Mendelism is not the real method or path of organic Evolution, but at best only an important side-track. This is not intended as a reflection on the science of Genetics, but only to place it in a proper perspective in the whole field of organic Evolution.

Having ruled out Mendelism, can we accept De Vries's Mutation as the ordinary method of creative Evolution? Mutation takes place when specific or varietal characters appear, not as the result of a slow age-long summation of small variations, but at one bound, with a great leap of one generation to the next. An individual of species X produces offspring which constitute a stable variety or a new species Y. De Vries saw this happening in the case of cultivated *Œnothera lamarckiana* growing wild in a potato-field at Hilversum in Holland. Other instances have been observed by other investigators. It is objected that De Vries's Œnothera was perhaps a cultivated artificial hybrid, with mixed qualities, like the *ab*'s of the Mendelians, and that all he observed was the emergence of pure qualities from this mixture; in other words, not the emergence of new characters but the setting free and unmasking of concealed or latent characters already existing. Other criticisms also have been levelled at the Mutation theory which it is not necessary for our purpose to consider here. In spite of these criticisms it is practically certain that mutations do take place in the course of Evolution. But while they almost certainly happen on special occasions, they are not common and do not constitute the ordinary method of organic Evolution. On rare occasions there is a *saltus*, a creative leap forward from one generation to another. A species having long balanced itself precariously on the edge of a great change suddenly makes the jump, secures a foothold on the edge of the other side, and marks the beginning of a new variety or species. But it can at best only be an exceptional if not rare effort on the part of Nature. These sudden long jumps can only be very occasional, and not the normal course or procedure in the origin of species. Otherwise we would certainly see more of them, and they would not be the subject of doubt or dispute. The rarity of their observation points to the rarity of their occurrence. And they must be largely confined to cultivated artificial species or varieties which are more unstable and violently variable than natural species or varieties. Mutation in wild nature is an occasional and exceptional occurrence, and is not the ordinary procedure of Evolution.

Having thus ruled out both Mendelism and De Vries's

Mutation as the usual method of creative Evolution, we now come back to the earlier Germ-cell theory of Weismann, who initiated it and through it the second phase of Darwinism, and thus became, and still remains, the second most important figure in the history of Darwinism. His great and essential service consisted in this, that he found the real source of Evolution in the inner factor of Variation, and that he traced this factor to its seat in the germ-cells of the organism. Not the outward mechanical struggle and clash of organisms, but the penetralia of their deeply hidden and sheltered germ-cells were the mysterious, spontaneous, independent and original source of all organic development and of the origin of species. Of course this theory became possible only by reason of the rapid advance in the knowledge of the cells, and especially of the part they play in reproduction. But on the basis of that new knowledge the theory became quite simple and indeed inevitable. The body-cells of advanced organisms have no part or lot in reproduction, and the seat of all organic variations must therefore be looked for in the reproductive cells of the parents. All organic progress was thus traced back to the inmost nature of the organism itself, and not to the environment or any mere external factor. This is the essential truth in the hypothesis of Weismann, and this constitutes his real and lasting contribution to the theory of Evolution. The mysterious Variation which forms the inner factor of Evolution has its seat and source in the fructified ovum or germ-cell from which the new life begins. There and nowhere else take place the great play and inter-play of forces, tendencies and influences which shape the destinies of life in organic development. This is not the whole story, but it is important; it is indeed fundamental.

Weismann drew a sharp distinction between the individual and the race, between the body-cells which constitute the one and the germ-cells which are the carriers of the other. According to him the race or species is continued unbroken in the substance of the germ-cells, which flow on as a continuous stream from one generation to the next. From these racial germ-cells are differentiated the body-cells in the individual life, both in its ante-natal and post-natal stages. After the differentiation has taken place in the fructified ovum, there is, according to him, practically no connection between the germ-cells and the resulting body-cells which build up the individual, except in so far as the former are nourished through the latter. The individual becomes separated from the race factor, and becomes an independent growth from it, becomes, so to say, an excrescence or epiphyte on the race, which continues in the germ-cells uninfluenced by the fate or the development of the individual. This complete severance and independence of the individual from the germinal constitution from which it has sprung is a distinctive tenet of Weismannism. It embodies a profound truth, which we recognise in the freedom and independence of individuality. But at the same time it makes the severance of the racial and individual elements in the whole too great, and it ignores important reciprocal influences between them which maintain a certain balance between individual and racial development. To these points we shall have occasion to return. Here it is instructive to note that for Weismann the sharp distinction between the individual and the germ-cells, from which it sprang and which it carries forward for the race, was based on his view of the nature and constitution of these germ-cells. These cells contained the hereditary constitution of the race or species, and in so far registered the past, and made the past an operative factor in the present. They also embodied the mechanism of variation and thus linked the future with the past in the continuity of the race. In a way, therefore, the germ-cells, uninfluenced by the ephemeral and accidental influences of the individual life, contained in their wonderful constitution not only the present but also the past and in a measure the future of the race. They were eternal, self-contained units, carrying their future and their past in themselves, uninfluenced by the accidents of their environment. The individual was a mere bit of bread cast on the waters of destiny, to be lost utterly, or to be found after many days. But the past and the future of the race dwelt sublime and secure in the eternal sanctuary of the germ-cell.

Such was the great Weismann conception, which in effect largely withdrew creative Evolution from the arena of external conflict and the mechanical struggle for existence, and located its origins in the secluded depths of the inner world of the germ-cell. And this great conception was based on Weismann's view of the mechanism of the germ-cell, on which a great deal of light has since been thrown by experimental research and observation.

Without going into details we may just note that the chromosomes of the dividing nucleus have been identified as on the whole the carriers of the hereditary characters of organisms; these characters have to some extent been correlated with distinct chromosomes, and the number, shape, size and other differences of chromosomes in the nucleus of the germ-cell are therefore taken to be the physical basis of the characters which distinguish the species. It has been found necessary to go further and to assume in the chromosomes themselves active elements or factors or genes which are productive of organic characters. These researches and speculations are still in their initial stages, but they are important and have this advantage, that the results of intercrossing and hybridising in producing a change of characters can be studied in conjunction with the simultaneous change in number and form of chromosomes. In the prosecution of experimental Evolution the parallelism of cell structure and of variation in organic characters thus supplies a double weapon of attack.

While the germ-cell as the mechanism of heredity is easily understood, the question still remains how it operates as the sole and independent cause of Variation. The intermixture of chromosomes from two separate individuals in sexual reproduction, and the changes in the chromosome contents of the reproductive cells in their previous meiotic division, undoubtedly provide the occasion for a great intermixture of parental elements and are thus potent sources of Variation. But Variation operates even apart from and in the absence of sexual reproduction and the related meiotic divisions of the germ-cells. And the question remains whether the individual life is, in fact, so isolated from the germ-cell that it has no influence on the latter and the resulting offspring. On this isolation Weismann was particularly insistent, and in the popular mind his teaching is identified with the doctrine that acquired characters are not transmissible. The principle of the non-transmissibility of organic modifications (as above defined) rests on empirical experience, as no clear and indisputable case of the passing of such individual modifications to the offspring has been recorded or observed. Weismann's germ-cell theory was intended to supply the scientific basis for this negative result; but in the end he so completely isolated the germ-cell from the rest of the individual organism that he came to consider it practically impossible that modifications could become hereditary, or that somatic cells could in any way, except through nourishment, influence the germ-cells.

There can be little doubt that in adopting this extreme standpoint Weismann went too far. He not only cut clean away from the Darwinian tradition, but also, in fact, made it impossible to understand the double fact of progress and adaptation; in other words, to understand how the experience of the race, which after all is only accumulated individual experience, helps to promote development, and to mould it in congruity with the environment. Unless the "trial and error" experiments of individuals produce some racial result; if, in other words, every individual throughout the ages has to begin to learn once more at the beginning, organic progress becomes unintelligible, if not impossible. The extreme isolation and independence which Weismann attributed to the germ-cell therefore led to a further hypothesis intended to give the individual some sort of indirect influence in shaping racial evolution. He assumed that a struggle for existence took place among the elements or genes inside the nucleus of the germ-cell for the food that came from the body-cells, that Natural Selection was thus already at work inside the germ-cell, and that it was the vigorous, well-fed surviving genes that shaped the course of the resulting variation in the direction to which the individual had thus contributed. In this way the body-cells and the individual modifications of the parent might have some vague and indirect influence on the germ-cells and their offspring. This arbitrary and unsatisfactory hypothesis has found no favour and probably amounts to no more than a confession of failure on the part of Weismann to maintain his doctrine in its extreme form. To transfer the venue of the struggle of existence from an arena where we can watch and observe it among organisms to the inner arcana of the germ-cells, where it is beyond observation and where its operation, if any, is pure guesswork, is not a helpful hypothesis, and can only be a last desperate resort of a theory in distress. Weismann no doubt felt the difficulty keenly, but he saw no way out of it, and his hypothesis of Germinal Selection was no way out.

The dilemma is indeed a most formidable one, not only for Weismann but also for all current views of Darwinism. On the one hand, there is the negative evidence, the absence of any clear and incontrovertible case where mere individual modifications have been transmitted to offspring. On the other hand, there are the very numerous cases where the disappearance of certain characters can only be satisfactorily explained on the assumption that modifications due to disuse of an organ have become hereditary. Again, there are the still more numerous cases where parts of the body have been constantly used in certain ways and have finally become specialised organs with which animals are now born ready-made. There is also the class of cases mentioned by Herbert Spencer in his controversy with Weismann and never satisfactorily answered, where, for instance, the sensitiveness of the finger or tongue (now hereditary) is compared with the much smaller sensitiveness of the back or other parts of the body, which have never been used as an organ of touch or taste. Above all, there is the difficulty, one might almost say the impossibility, of understanding organic Evolution, if its advance depends upon mere fortuitous variations in reproduction, and remains uninfluenced by the work, the experience, the learning through trial and error and the consequent modifications of the individuals which compose a race or species. While it is admitted and intelligible that mere artificial and singular modifications, such as cutting off the tails of dogs or sheep continuously for thousands of years, will have no germinal and no hereditary effect, the case may apparently be quite different with modifications which are due to the frequent or constant activity of the animal, and which register the routine of its life. Such modifications are far more intimate to the animal organism, and may in the course of time produce such a deep impression on the body-cells as to penetrate to and reach even the germ-cells, and register a change there which leads thereafter to hereditary and apparently spontaneous variation.

Apart from Weismann's extreme doctrine of germinal isolation, which even he by implication appears to have found untenable, there is nothing in principle directly negativing such an assumption, and it does render intelligible the progressive evolution and specialisation of bodily organs which on any other assumption it is most difficult to understand. The absence of direct experimental evidence in support of this view is not a fatal objection. The laboratory of Nature is very different from that of experimental research. Life has not been made in the latter but was made in the former. The slow intimate operations extending over thousands and even millions of years, such as brought about most of the organic species of which we know, are not on a par with our latter-day researches in experimental evolution. With all our chemical knowledge we can yet never hope to rival in our laboratories the results which Nature has through the countless ages achieved, say, in the crucible of the geological record. Still less can we hope to achieve through biological experiments in the laboratory what her silent processes have amounted to through millions of years.

If we set aside this negative and really irrelevant evidence, and also reject Weismann's extreme doctrine of germinal isolation, we find nothing in theory or fact to preclude us from viewing modifications as having an influence through more or less long biological periods on the germ-cells. On the hypothesis of the "field" which we have found useful before, we may consider these somatic modifications as in the first instance influencing the field of the germ-cells, and only later and in the course of time becoming incorporated from the field into the hereditary structure of the germ-cell.

We come thus in effect to look upon modifications as partly the material from which variations have been formed. Modifications due to constant use or disuse, or to permanent changes in the conditions of physical environment, influence in the first instance the field of the germ-cell, and are thus the earlier phase of the later hereditary structural variations. In fact we may say that modifications are to variations what variations are to specific characters. Throughout organic Nature we find this grand calculus at work, adding up and conserving whatever in the experience and development of the individual is of survival value to the race, and carrying on this organic summation with a fineness and delicacy surpassing that of any mere mechanical calculus. What is not incorporated into the hereditary structure remains conserved in the invisible "field" until it is finally accentuated enough to become so incorporated. Nothing of value is lost—traces and residua of organic reactions, reflexes and tropisms, instincts and intelligence, all are conserved or registered in the field until in the lapse of time they are ready to become part of the physical structure. There is no reason, except our ignorance of the facts, why modifications should not thus to a large extent be the conditions precedent of variations. Only in this way can we explain why the trend of Variation is on the whole in harmony with the experience and the past of animate

Nature, why Evolution makes steps in advance on the road on which it is already moving, instead of making incalculable twists and turns, as it might do if its course was merely dependent on purely accidental, arbitrary and unmotivated variations.

That modifications of a certain intimate bodily character, and continued through many generations, may in the end influence the germ-cells and even modify their hereditary structure is easier for us to appreciate than it was for Weismann. It is only recently that we have learnt to understand the important functions which the hormones given off by the ductless glands perform in the regulation and balance of our whole animal economy. We now know that the germ-cells, so far from being independent of the developed system of body-cells, have even apart from their reproductive functions a most intimate regulative effect in co-ordinating the functioning of the bodily system as a whole. If there is this open door between them, there is no reason why there may not be the reverse influence of the body-cells on the germ-cells.[1]

This question of the way in which non-hereditary modifications are conserved brings us to another difficulty which Evolutionists have found it very hard to explain on the accepted Darwinian principles. I refer to the natural selection of small variations. How can small variations be selected and conserved in the struggle of existence until they are marked enough to become specific? To begin

---

[1] From this point of view the recent experiments of Professor Pavlov on the associative memory of white mice are interesting. An electric bell was rung while the mice were feeding. It was found that a firm association was built up after this process had been repeated 300 times; that is to say, after that the mice looked for their food whenever the bell was rung. For the children of these mice a less arduous lesson was necessary: after 150 rings the association was established. For the grandchildren only 30 rings were necessary; while for the great-grandchildren (the third filial generation) only five rings were necessary to establish the association. In other words, the acquired experience of the parents made the acquisition of similar experience progressively easier for their offspring. The results of these experiments are not generally accepted. If these experiments are widely corroborated they will throw a new and most important light on the nature of Evolution as progressive facilitation of experience; in other words, on the hereditary character of educability or psychic experience.

with, they are so small that it is difficult to understand that they have any survival value at all. Take an organ which is being differentiated from the rest of the body-cells. At the beginning any variation must be utterly insignificant and practically valueless in the struggle for life, and Natural Selection has really nothing to work on. How then could an animal with such a minute variation be selected as being more adapted to its environment? It is this awkward question which has led to the hypothesis that very marked varieties or mutations alone are selected. Various more or less ingenious attempts have been made to answer this question, but to my mind they are all more or less unsatisfactory. The result is that we cannot understand how the Darwinian machinery of Natural Selection is set in motion in any particular case. Once individuals with marked specific or varietal differences exist in superabundance, we can understand why the struggle for existence between them will take place and Natural Selection become operative. But on Darwinian principles as ordinarily understood these marked differences between individuals can arise only from a prior selection as between variations so minute that there is apparently nothing sufficiently substantive for Natural Selection to work on. In other words, Natural Selection will move all right when once set in motion, but Darwinism fails to set it in motion.

In my view the difficulty can be satisfactorily removed only by the principle of Holism, as I shall just now proceed to explain. In the meantime, however, I wish to point out how my suggestion that the modifications influence the field of germ-cells and prepare the way for variations can prove helpful to Darwinism in its plight. According to that suggestion the small initial variation does not stand by itself, and on its own merits, so to speak. It appears powerfully supported in the struggle for existence. Individual use and practice for very many generations are on its side. It does not appear as a stray, helpless infant in a hostile world. It appears in a friendly, one might say, in a prepared universe. It has a stalwart nurse in the use and routine of the individual in whom it appears. It is protected, shielded and in its struggle reinforced, by this constant use and routine. A small variation in the direction of a nascent organ, for instance, finds itself in line with the traditional use of generations of individuals which powerfully support it in the struggle with contrary variations. Under the shelter of this use it develops and beats its competitors, until in the end it can fend for itself and engage in the struggle on its own account.

This explanation applies not only to variations in developing organs which are supported by use and practice on the part of a long line of individuals. It applies also to cases where permanent changes in the physical conditions impress themselves continuously on the organism. The growth forms of plants, for instance, under particular ecological conditions are such as almost to render necessary the view that ecological modifications, due to the direct, silent, long-continued pressure of the environment, finally become variations. The sameness or close resemblance of the growth forms under the same physical conditions, as seen, for instance, in the general characters of formations and associations in the vegetable kingdom, are probably in a measure due to the age-long operation of ecological factors which have impressed themselves on plant development and have produced modifications which finally have become variations.[1] The resulting general features of formations and

[1] While this book was going through the press I was much interested to see this view corroborated by certain observations of Professor F. O. Bower in *Evolution in the Light of Modern Knowledge* (p. 206). After discussing the evolutionary structures of ferns he continues: "It would seem a natural interpretation of the facts that the characters (under discussion), acquired by a direct impress upon a succession of individual lives, should have been imposed hereditarily upon each race. Naturally the reply may be made that probably mutations favourable to the perpetuation of the imposed character may have made that character permanent. If we grant that, do we not thereby simply admit that the distinction between fluctuating variations and mutations is not absolute? In other words, that fluctuating variations repeatedly imposed upon successive generations are liable to become mutations? It is difficult to see any other rational explanation of the wide-reaching facts of homoplastic adaptation, such as are shown with exceptional profusion in the ancient class of the ferns, and are evident in plants at large." (In this quotation fluctuating

associations are no doubt in part due to Natural Selection, but in part the physical environment has probably exercised a direct pressure all its own and produced an effect which has powerfully reinforced the results of Natural Selection. The hereditary variation ultimately appears, but it does so not accidentally or from the blue, but from the long-continued stimulus of environmental conditions which have influenced and affected the field of the germ-cell.

While some variations thus have their roots in the traditional use and practice of individuals or in the conditions of the physical environment, and can survive under the protection thus afforded them, many variations cannot be thus accounted for, and probably originate in what appears to us as a spontaneous, independent, more or less sudden and accidental manner. The mode of their selection and survival has still to be accounted for. Before doing so it is advisable to mention a third set of difficulties which Darwinism encounters in its explanation of organic Evolution. I refer to the phenomena of co-ordination and co-adaptation of organs and characters which it is almost impossible to account for satisfactorily on orthodox Darwinian lines.

I have hitherto spoken of variations as if they came singly in the evolution of organisms. But they do appear but rarely as single units. Generally they appear in associated groups. A small variation is generally found to be accompanied by a number of still smaller associated variations. If an organ varies, the associated muscles, nerves and other body-cells undergo a corresponding variation. The evolution of the horns of a wild beast, for instance, means minor and consequential adjustments to its head, its neck, its muscular system, the development of the forepart of the body, and its relation to the back parts, as well as to many other parts and details of its body. And

variations correspond to what has above been called modifications, while mutations correspond to what has been called variations.) The experiments of Kammerer and Durkhen on animals and plants would seem to tend in the same direction. But they require further corroboration.

when we come to consider the question, already so difficult, of the selection of a small variation in respect of such a horn, we are confronted with the still more hopeless difficulty of having at the same time to account for many other minor correlated variations, each of which has to be selected. Besides this, there is their joint and associated use or functioning which has also to be accounted for as a factor in their selection. We are obviously throwing a weight on the principle of Natural Selection which is more than it can bear. It is being arbitrarily and artificially applied far beyond the area of its natural and proper application. And here it is where Natural Selection breaks down completely. The whole body is a system of co-ordinated structures and functions, and its origin and development can be represented only as a complex movement forward in time of a mass of associated variations which have resulted in the most marvellous co-adaptation of structures and co-ordinated functions. Before the problem of this complex yet orderly evolution, Natural Selection stands baffled. It can deal with individuals and markedly formed and developed characters, but not with their delicately adjusted and associated infinitesimals.

The fault, however, lies not so much with Natural Selection, as with our fundamental organic conceptions. Our crude uncritical mechanistic conceptions are the real source of the difficulty, and Holism appears to me to be the way out. The root of the error lies in our disregard of the individual organism as a living whole, and in our attempt to isolate characters from this whole and study them separately, as if they were mere mechanical components of this whole. The fatal mistake involved in this procedure has already been fully exposed in previous chapters. The whole is not a mechanical aggregate indifferent to and without influence on its parts. It is itself an active factor in controlling and shaping the functions of its parts. The parts bear the impress of its directive influence, without and apart from which it is vain to speculate on their characters and their activities. Whereas mechanical action is isolable and additive, so that the total activities of a system are represented by the sum of all the individual activities, the situation is entirely different in the case of a living whole. Here all action, as we have seen, is holistic, not only that of the whole itself, but also that of the parts. The stamp of Holism is impressed on the activities of the parts no less than on the individual whole itself. The individual and its parts are reciprocally means and end to one another; neither is merely self-regarding, but each supports the other in the moving dynamic equilibrium which is called life. And so it happens that the central control of the whole also maintains and assists the parts, and the functions of the parts are ever directed towards the conservation and fulfilment of the whole. With this conception of living unity and holistic action in an organism before us, let us try once more to read the riddle of Variation and Natural Selection as the twin factors in Evolution.

In the first place we realise that each individual organism is a unitary system whose inmost nature is its own balanced self-maintenance and self-development as a whole. Heredity is but the expression of this self-conservative character. The organism both as structure and field, while carrying with it the past which is its expressed self, also carries with it the still unrealised future which flows organically from that past, and it maintains a living, moving harmony between the two; its presently existing self is the more or less harmonious realisation of the organic unity of its past and its future in its present. Variations arise as the tentacles it throws out under environmental stimulation towards the future, a stretching of hands dimly and unconsciously towards future adjustment, welfare and betterment. These variations, while apparently accidental and uncontrolled, arise from the stimulus of the environment and are under the central control of the organism as a whole.

Let us for a moment consider the appearance of a small variation. It is really neither spontaneous nor accidental. It is the expression of the moving, developing organism as a whole in a particular direction. It is normally conditioned by what has gone before in the history of the organism and is really of a piece with the organism as a whole. Nor does it as a rule appear alone. The organism as a whole is on the march, and while the variation may be the first and most significant indication of the inner movement, the advance is not confined to a single point, but is represented by a curve of progress on which other minor advances are registered at the same time. Thus variation A when closely scanned will be seen to be really more like $A + b + c + d$, where b, c and d represent minor variations which adjust A in various respects to the organism. The apparently isolated variation is seen to be what it really is, an advance of the organism as a whole in a particular direction, a holistic as distinguished from a singular and mechanical variation or change. Mechanical analogies may assist us to understand to some extent what happens. A mechanical system of a given number of elements in equilibrium is given a push or blow with a certain force in a certain direction. When it has recovered from the push or blow and is in equilibrium once more, it will be found that the change is not merely in the direction in which the force was applied, but that all the other elements have also been affected and have undergone adjustments in order to achieve the new equilibrium. The same happens, only much more intensely and intimately and organically, in the case of a change in a living whole. Variation A necessarily involves a number of collateral adjustments which are dependent on A, and are not independently originated or conserved. In other words, holistic variation or variation of a whole in any particular respect is the cause and carrier of minor variations which are not independently selected or conserved, and for which Natural Selection need not, therefore, be called into action, It is really the whole which does the "selection" in the exercise of its central control. We may call it a case of Holistic Selection as distinguished from external Natural Selection. Variation A of the whole, which is the expression of an inner urge of the whole and is therefore supported by the whole, carries with it the minor and consequential adjustments involved in variations *b, c* and *d*.

This explains one of the main difficulties which we encountered above—the question, that is to say, of the selective co-ordination of subsidiary adjustments. But the main difficulty remains how variation A itself is selected after its appearance. How is the main small variation, perhaps too insignificant for Natural Selection to get a grip on, selected and conserved in the holistic system? If it were a mere accidental appearance, with nothing more behind it, it might be a toss up whether it is saved or lost, and generally it is lost. With the prodigality of life itself, organic changes are scattered broadcast like seeds, and most of them, with nothing particular in the urge of the organism behind them to give them continuous momentum, perish as soon as they are born. But some are in a different position; they are in the main direction of development, they are on the road, so to say, on which the organism is travelling; they have the whole weight of the organism behind them; they are nursed and cared for, figuratively speaking; and in the end they survive. Once more a case of Holistic Selection as distinct from Natural Selection. And sometimes in these cases, as we have seen, the organism has long before the appearance of the variation begun to move in its direction. The functioning of the organism has anticipated its future structure. It has for many generations devoted a part of itself to a particular use; the part has in consequence undergone modification; from an undifferentiated system of cells it has been modified in certain respects so as to anticipate an organ. When finally in the course of time this modification is superseded by and merged into an organic variation, it is in direct harmony with the needs and the practice of the organism as a whole; the practice continues along with the variation and becomes accentuated, the pressure of the needs of the organism is behind the variation and probably increases; and the variation, covered by the habitual practice of the organism, and urged forward by the organic needs, makes headway and has a fair chance of survival.

It has a distinct advantage; the dice are loaded in its favour by the nature, pressure and practice of the organism as a whole. These forces behind it are probably strong enough to keep it going, though only at the very slow pace at which all biological Evolution moves. Eventually, when it has developed enough to add a sensible measure of strength to the parent organism, it will reward its parent for its secular support, it will join forces with it, and fight a victorious battle against its competitors. At this stage the belated force of Natural Selection has arrived on the scene. But not earlier, the earlier phases having depended on what I call Holistic Selection.

The Holistic Selection which acts within each organism in respect of its parts *inter se* is essentially different from the Natural Selection which operates between different organisms, which is more appropriately called the struggle for existence. Holistic Selection is much more subtle in its operation, and is much more social and friendly in its activity; it puts the inner resources of the organism behind the promising variation, however weak and feeble it may be in comparison with other characters, and makes it win through powerful backing rather than through the ruthless scrapping of the less desirable variations. In the organism the battle is not always to the strong, nor is the struggle an unregulated scrimmage in which the most virile survive. The whole is all the time on the scene as an active friendly arbiter and regulator, and its favours go to those variations which are along the road of its own development, efficiency and perfection.

The continuous Holistic Selection of small variations may be compared to the survival of obsolete organs in an organism. Both are carried forward by the organism as a whole, perhaps for millions of years, without being in either case directly useful to the organism. The whole, if one may say so, takes long views, both into the future and into the past; and mere considerations of present utility do not weigh very heavily with it. It carries its infant variation with it in the same way that it carries the aged and dying members or atrophying organs. Both are borne along, covered and shielded by the main characters of the organism. From the point of view of survival value, as from so many other points of view, the whole is more important than any of its parts. And so it comes that the organism is a most complicated system, a present living unity embodying its faraway past no less than its dim distant future. The whole controls, guides and conserves all. The fate of any particular part, considered by itself and on its own merits, would be an inexplicable mystery, and might be expected to be the very opposite of what happens in practice. When, however, it is viewed from its position and function in the whole, the mystery is explained; we see how different the laws of life are from the laws of mechanics, and how wrong it is to apply mechanistic and atomistic conceptions in a region where Holism prevails.

To understand how a small variation is favoured, we may represent an organism as a moving developing equilibrium, which is never perfectly adjusted because it has a persistent slight overbalance in the direction of development. Complete equilibrium is never attained, and would be fatal if it were attained, as it would mean stagnation, atrophy and death. And so the overbalance in a certain direction or with a definite orientation continues indefinitely, and all small developments and adjustments and "variations" which have that specific orientation have the momentum of the whole behind them and tend to survive and grow while others in other directions are dropped and discarded. One may accordingly say that in each case "the whole" is a co-worker with its small variations which will eventually be useful; that as an active factor its influence is on the side of such small variations, and that with this inner nurture and support these small variations are practically independent of external support for their survival and steady evolution.

The activity of the whole is seen not only in the maintenance and evolution of the small variation and all the subordinate adjustments that go with it, but also and especially in all the innumerable co-ordinations and co-adaptations in structure and function which constitute a living organism. I believe it is generally admitted that this phenomenon of organic co-ordination is one which cannot be satisfactorily explained on mechanical principles. The functioning of an animal as a whole has something unique about it, and the term "whole" in this connection is no mere phrase but a fact of vital significance. We have already considered the matter fully in Chapter VI. Here we shall only add that to suppose that Natural Selection has not only brought about the separate organs of animals and their functions, but also accounts satisfactorily for their adjustments to each other and their co-ordinated activities in the animal behaviour, is to suppose what certainly has never been and cannot be explained in detail, and what probably is in conflict with the facts of development. Intelligent and purposive action of a human or other animal cannot be explained on mechanical principles; nor can instinctive action, not even reflex or organic activities and functions below the level of instinct or intelligence. An animal—even of the lowest type—makes an unconscious effort to catch food or beat an enemy, and in the process performs a large number of acts which are all effectively co-ordinated towards the attainment of its object. No mechanical explanation of this process of co-ordinated movements has ever been given. The animal acts as a whole, with a unity and effectiveness of action which is no mere mechanical composition of its movements. The concept of the whole is the only category that will explain such unity, and we have seen good reason in previous chapters to go further and to infer that Holism is not merely a category or concept, but a fact and a factor of far-reaching significance. Co-ordination and co-adaptation in organic structure and behaviour cannot be explained on any other ground.

So far we have considered Holism as creative of variations; and as regulating and co-ordinating groups of actualised variations and organic characters generally. But this by no means exhausts the function of Holism in organic development. It is not only productive of variation, it is just as much repressive of variation. Holism is as often inhibitive as creative; it keeps back certain elements at the same time that it pushes forward others, and in this way secures a balanced movement and progress of the organic whole. When, for instance, the form and characteristics of a gorilla are compared with the human type it becomes clear that in the human evolution certain tendencies have been held definitely in check, and the utter caricature in appearance, which would have resulted from unrestrained development, has thus been prevented. Nobody who ignores this negative aspect of Evolution can possibly understand the present forms of animals, compared with their living or fossil affiliations. Tendencies to variation, which were realised in the case of Neanderthal man, have been more or less severely repressed in the present human races. If there had been unrestrained evolution of all potential variations, the results would have been truly dreadful in their grotesqueness. In fact we find at work in organic Evolution an influence not unlike that which at a much later stage we recognise as the ethical control of feelings, impulses and instinctive movements of an undesirable character. The whole in personality, the whole in its ethical flowering in the human, means not only expression of certain moral qualities, but also and equally repression of others. Elements and tendencies which we find strongly operative in our instinctive or organic nature we have to keep in check, to hold down severely, and to prevent from emergence in our characters as a whole. This is the very essence of Holism in its mature ethical development. There is something very similar and equally fundamental in the activity of Holism in the earlier purely organic phases of Evolution. In any individual organism the whole is in control, pushing forward some tendencies and keeping back others, expressing some variations and repressing others, and through all maintaining a mobile equilibrium of all the elements, positive and negative, that are uniquely blended in the individual. Thus it is that if we wish to understand the details of organic Evolution in any particular case we should look for the repressions no less than for the variations; it is the combination of the two which constitutes Evolution.

I shall no doubt be asked what experimental verification there is for the holistic view of Evolution here set forth. My answer is to repeat what I have already said: that natural Evolution as distinguished from experimental Evolution is a process, not of the hour or the day, but of geological time, and that the results, matured and consolidated through immemorial periods, cannot be repeated or rehearsed by short-dated laboratory experiments, conducted too under conditions very different from those of Nature. These experiments, however valuable and instructive in affording subsidiary clues and hints of the natural process, do not by any means exhaust or even seriously affect the real problem of creative Evolution; and a correct view of Evolution must have regard to this difference and be based on an intelligent appreciation of the natural processes rather than on the very limited data yielded by our laboratory experiments. There is no doubt that experimental Evolution has, through its unavoidable limitations, greatly blurred the great Darwinian vision of organic Evolution, and instead of making us more fully realise its truth and effectiveness and grandeur as a whole, has tended to deflect our attention to particular problems which are special and limited enough to be capable of laboratory treatment. The special and exceptional cases of Mutation and Hybridisation come to be looked upon as covering the entire process of organic Evolution. My endeavour in this chapter has been, through a re-examination of the position thus created, to explore and reconnoitre a way back to the broader and wider view of Evolution. And in doing so I have sought the assistance of a concept which we have found at work, not only in organic Evolution, but in all organic structures and processes and even, to a limited extent, in inorganic Nature itself. I shall now briefly summarise the results we have reached in this chapter in order thus to see how they bear on the wide Darwinian conception from which we started.

The relative importance of the internal and external factors in Evolution has materially altered since Darwin's time. Variation has become much more important than Natural Selection, not only in biological studies and experimental researches, but also in our view of it as an operative factor in organic Evolution. While remaining a substantial and important factor Natural Selection has yielded pride of place to Variation. The factor of intense struggle and competition in Nature on which Darwin, following the Malthusian clue, laid so much stress is now seen not only to have less importance relatively, but also to bear a somewhat different character from what it had in Darwin's view. The struggle for existence is, like Mutation, an exceptional and not the usual procedure of organic Nature. This world is at bottom a friendly universe, in which organised tolerant co-existence is the rule and destructive warfare the exception, resorted to only when the balance of Nature is seriously disturbed. Normally Natural Selection takes the form of comradeship, of social co-operation and mutual help. Normally also the organic struggle is very much in abeyance, and the silent, effortless, constant pressure of the physical and organic environment exercises a very powerful influence. The young science of Ecology has been built up since Darwin's time and is based on the recognition of this fact, that, in addition to the operation of Natural Selection, the environment has a silent, assimilative, transformative influence of a very profound and enduring character on all organic life. In the subtle ways of Nature, sun and earth, night and day, and all the things of earth and air and sea mingle silently with life, sink into it and become part of its structure. And in response to this profound stimulus life grows and evolves, the lesser whole in harmony with the greater whole of Nature.

The interaction between the inner and the outer factors in Evolution is far more close and subtle than one would infer from Darwinism, in either its earlier or its later (Weismann) form. It is not merely a case of one factor creating variations, and another eliminating some of these creations and leaving free the rest, which are then said to be selected for perpetuation. The inner creative factor in a measure acts directly under the stimulus of the external factor, and the variations which emerge are the result of this intimate interaction. The isolation of the inner from the outer factor, which was so much emphasised by Weismann, is, in spite of its apparent agreement with observation, really a mistaken assumption, based on the neglect of the factor of time in Evolution. Environment is a great stimulus of variation, and even more so is the somatic organism itself, which is closer to the germ-cell than the environment.

We can understand the process of organic Evolution only if we assume that, deeply as the germ-cell carriers of Variation are hid from external contacts, they are not completely or for ever isolated therefrom; that changes due to habitual behaviour or to environmental or ecological pressure affect the " field " of the germ-cells, and if sufficiently long-continued and intense, sooner or later penetrate the structures of these germ-cells, and stimulate and set in motion the internal factor of Variation. The response comes back in a crisis of variation or mutation which permanently alters the internal hereditary structure. In these cases the inner and outer factors of Evolution do not operate independently and by opposed and contrasted methods; they collaborate in the closest manner as the stimulus and response which we find distinctive of all organic action. From this external factor, which operates as a stimulus of organic variation, we have to distinguish Darwin's Natural Selection, which is another external factor operative not in connection with the stimulation of variations, but in connection with their subsequent elimination or destruction. The external factors in Evolution are therefore according to this view twofold: the environmental or ecological factor which to some extent influences or induces variation, and the factor of organic struggle which sets in motion the warfare among organisms for the limited goods of life, which Darwin called Natural Selection.

But it is only in certain classes of cases that the " use " factor or the external or ecological stimulus of variation comes into action; in others the stimulus of variation is entirely internal, and must be found in the fresh mixture and readjustment of the chromatin elements of the germ-cell nucleus at certain critical stages in the evolutionary process, such as in the sexual reproduction of some organisms, or the endomixis and rejuvenescence which occur at certain stages in others.

This internal factor in Variation and Evolution was stressed, and rightly stressed, by Weismann, and has supplied a suggestive clue for the researches in Genetics which have been conducted since his day. But the view of this factor as purely mechanical has led to great difficulties in detail, and has made it impossible to understand the process of organic Evolution as a whole. I have therefore endeavoured to stress the contrary view of this inner factor, and to show that it is holistic in character and operation, that it thus solves the difficulties which the mechanistic hypothesis has created for itself, and that it leads to a reconciliation of the two factors operative in Evolution. Holism has thus once more, though in a way different from that envisaged by Darwin, brought us back to the great Darwinian vision of universal adaptation.

But Holism has done more; it has enabled us to realise the pervasive creative unity which makes all the diverse elements of existence the co-operative members and inhabitants of an essentially friendly universe. Operating as the inner creative factor at the heart of things, it has led to the evolution of a universe in which all the factors and products, organic and inorganic alike, are not alien to and destructive of each other, but are capable of mutual adaptation, and adjustment, just because they own a common origin and have an indisputable, though often scarcely recognisable, family relationship. This is not to assume a Pre-established Harmony, which would be as great a mistake in one direction as the contrary and more usual mechanistic assumption—that universal adaptation and organic co-ordination are in effect the accidental results of utterly unconnected factors—would be in the other direction. The true conception not only for philosophy but also for science is that of parts in a whole. It is the high task of science to explore the mechanisms of adaptation and variation in all their details, and to pursue at all costs the chemistry and physics of the cell, of which we still know so little. But in doing so it must also explore the unifying, regulating, co-ordinating activity of the holistic factor, which even from a purely scientific point of view is just as important as the study of the special mechanisms. Above all, biological science must ever keep before itself the standpoint of the whole, without and apart from which all the details—so far from being recognised as being organic to each other—are mere loose meaningless items, like the sands of the seashore, utterly useless for the understanding of that unique unity which constitutes an organic individual. The whole is the ultimate category not only of organic explanation, but also of organic adaptation and evolution. And it is more than a category; as the creative factor of inner structural and functional control operative in all existence, it is the ultimate real in the universe and the creative source of all reality, whether organic or inorganic. Nay, more: Holism is also creative of all values. Take the case of organic Beauty. It is undeniable that Beauty rests on a holistic basis. Beauty is essentially a product of Holism and is inexplicable apart from it. Beauty is of the whole; Beauty is a relation of parts in a whole, a blending of elements of form and colour, of foreground and background, of expression and suggestion, of structure and function, of structure and field, which is perceived and appreciated as harmonious and satisfying, according to laws which it is for Æsthetics to determine.

It may be a question how far the phenomena of repression in conjunction with expression in organic Evolution, of regulated development as a whole, of beauty and of similar phenomena can be properly subsumed under the Darwinian factors. Perhaps it is better to recognise that there is something wider and deeper at work in Evolution than the factors as found by Darwin and his successors, something of which those factors are themselves but an expression. The whole is itself an active factor, and its activity as such explains phenomena which it is difficult if not impossible to account for in any other way without very forced interpretations. The inner sources of wealth and beauty in Nature are inexhaustible, and they are poured forth with a lavish hand in the creative process of Evolution. Not merely survival values on Darwinian lines count; on the foundation of variations with survival value is raised a superstructure of development which far transcends that narrow basis. Mind in its marvellous human efflorescence rests no doubt on a basis of survival value; but how much more it is than that! The glories of art and literature, the peace of the mystic religious experience, the creative Ideals which lift this life beyond the limitations of its lowly origin —all these experiences and developments have built a new spiritual world on the humble foundations of survival values. In the kingdom of life is visibly arising its capital, the City of God. Apart from the great human development, beauty in Nature tells the same tale. The song of birds, with its primary appeal to sex, but with so infinitely much more in it than the mere sex-appeal; the glorious forms and colouring of birds and beasts and insects, which no doubt rise in and from the struggle for existence, but finally rise above it, and rob it of all its sordidness and drabness; above all, the wonder of plants and flowers, which were meant for the eye of birds and insects, but which contain so infinitely much more than the eye of bird or insect ever beheld or ever can behold—it is everywhere in Nature the same. Everywhere we see the great overplus of the whole. So little is asked; so much more is given. The female only asks for a sign to recognise the male, and to help her to select him and stick to him in preference to others. And for answer she gets an overpowering revelation of beauty out of all proportion to her modest request. The peahen has no discriminating understanding of the wondrous colouring of the peacock, which far transcends even our human powers; but in some inscrutable way something of an emotional nature in her takes it all in and is satisfied. It is deep calling unto deep; it is the whole appealing to the whole. There is evidently more in all this than the Darwinian factors can satisfactorily explain, and it would be both foolish and unscientific not to recognise this frankly. To me the conclusion of the matter is that the inexhaustible whole is itself at work, that Holism is an active factor interacting with the particular Darwinian factors, that not only its aim but also its output far exceeds the immediate present utilities and needs of organic Evolution, and that its bow is bent for the distant horizons, far beyond all human power of vision and understanding.

CHAPTER IX

MIND AS AN ORGAN OF WHOLES

Summary.—Mind is, after the atom and the cell, the third great fundamental structure of Holism. It is not itself a real whole, but a holistic structure, a holistic organ, especially of Personality which is a real whole.

Psychology treats mind in man and the higher animals as a factor or phenomenon by itself, and analyses it into various modes of activity, such as consciousness, attention, conception, feeling, emotion and will. In this work Mind is viewed from a different angle; it is a form of Holism and it is studied as a holistic structure, with a definite relation to other earlier holistic structures. It has, therefore, a much wider setting and performs more fundamental functions in the order of the universe than appears from Psychology.

Mind springs from two roots. In the first place, it is a continuation, on a much higher plane, of the system of organic regulation and co-ordination which characterises Holism in organisms.

Mind is thus the direct descendant of organic regulation and carries forward the same task. This is the universalising side of Mind, and appears in the conceptual-rational or reasoning activity, which co-ordinates and regulates all experience. Its physical basis is the brain and neural system, which is the central system of regulation and co-ordination in the body. It is thus the crowning phase of the regulative, co-ordinative process of Holism.

In the second place, Mind is a development of an "individual" aspect of Holism which already plays a subordinate part in organisms. In man it pushes to the front as conscious individuality or the Self of the Personality, and becomes as conspicuous a feature of developed Holism as regulative co-ordination, if not more so. This intense element of individuality is the principal novelty in the development of Mind, the real revolutionary departure from the prior system of regulative routine, and in Personality it culminates in a new order of wholes for the universe. Mind in its individual aspect is thus the chief means whereby organic Holism has developed into human Personality.

Mind is in some respects as old as life, but life outran it in the race of Evolution. Besides, Mind needed life as a nurse, and its full development has therefore had to wait for that of life. The

CHAP. IX MIND AS AN ORGAN OF WHOLES

extraordinary self-regulation of organisms must, therefore, not be put to the credit of Mind, which was essentially a later development of Holism.

Mind is traceable ultimately to inorganic affinities and organic selectivities. The "tension" of a body in disequilibrium gradually became covered with a vague "feeling" of discomfort, which had survival value; instead of remaining a passive state it became active as *ad-tension* or attention, and ultimately consciousness. Interest became appreciable. Simultaneously the active individual side of Mind developed as conation, seeking, experiment; and from this double basis Mind grew with phenomenal rapidity in the earlier species of the genus Homo.

The individual self-conscious conative Mind is rightly stressed by psychology as the Subject of experience, the Self, and ultimately the Personality. In the universal system of order this individual appears as a disturbing influence, as a rebel against that order. But the rebel fights his way to victory, achieves plasticity and freedom, and is released from the previous regular routine of Holism. Mind thus through its power of experience and knowledge comes to master its own conditions of life, to secure freedom and to control the regulative system into which it has been born. Freedom, plasticity, creativeness become the keynotes of the new order of Mind.

This is, however, only one side of mental evolution. *Pari passu* with this individual development the universalising conceptual-rational side of Mind also develops rapidly; its regulative Reason makes Mind a part of the universal order, and the individual and universal aspects of Mind mutually enrich and fructify each other, and on the level of human Personality result in the creation of a new ideal world of spiritual freedom. This union of the "individual" subjective Mind with the universal or rational Mind is possible because the individual Mind has itself arisen in the holistic regulative bosom. Pure individualism is a misleading abstraction; the individual becomes conscious of himself only in society and from knowing others like himself; his very capacity for conceptual experience results mostly from the use of the social instrument of language. The individual springs from universal Holism, and all his experience and knowledge ultimately tend towards the character of regulative order and universality. Thus knowledge assumes in the first instance the form of an empirical order, as a system of common sense. Gradually the discrepancies of this system are eliminated and knowledge approximates to science, to a scientific conceptual order, in which concepts and principles beyond empirical experience are assumed to underlie the world of experience. The scientific world-conception marks the triumph of the universal element in Mind, but only on the basis of the freedom and control which the individual mind has mainly achieved. Mind as an organ of the whole, while taking its place in the universal order, has emancipated itself from the earlier routine of regulation and has assumed creative control of its own conditions of life and development. Thus it creates its own environment in society, language, tradition, writing, literature, etc., instead of being dependent on an alien environment as on the organic level. Again, Mind frees itself from the intolerable burden of organic inheritance by inheriting merely the widest, most plastic capacity to learn, and letting the social environment and tradition carry on the onerous duty of recording the past. While the animal is hidebound with its own hereditary characters, the human Personality is free to acquire a vast experience in his individual life.

Mind has its conscious illuminated area and its subconscious "field." In this field the forgotten experience of the individual life as well as the physiological and racial inheritance exercises a powerful influence. It is this influence that proves decisive for our fundamental bias, our temperament, our point of view, and our individual outlook on persons and things. It is of an intensely holistic unanalysable character; it is even possible that our neural endowment carries with it more in the way of sensation and intuition than appears from the special senses; that the sensitive basis from which they have been differentiated has continued to develop *pari passu* with them and to-day forms a subtle holistic sense, a capacity of psychical sensing or intellectual intuition which explains our holistic sense of reality as well as other obscure phenomena, such as telepathy. So much for the influence of the past. The future also becomes a potent influence on Mind. Through its dual activity of conception and conation Mind forms "purposes" which envisage future situations in experience and make the future an operative factor in the present. Purpose marks the liberation of Mind from the domination of circumstances and indicates its free creative activity, away from the trammels of the present and the past. Through purpose Mind finally escapes from the house of bondage into the free realm of its own sovereignty. All through its great adventure its procedure is fundamentally holistic, and this can be shown by reference to the various activities of Mind as analysed by psychology. Free creative synthesis appears everywhere in mental functioning, and not least in the region of Metaphysics, Ethics, Art and Religion, which, however, fall outside the scope of this work.

IN previous chapters Mind has often been mentioned as a factor in Evolution. In all the references only the well-known meanings and activities of Mind have been assumed, and my procedure in making use of the factor of Mind in anticipation of its full discussion is therefore not so objectionable as it might appear from a purely theoretical point of view. The successive phases of the whole so telescope into each other that it is impossible to treat each phase in a water-tight compartment, and any attempt to do so would only result in a distorted view of the subject as a whole. In dealing with matter we had to anticipate the coming development of life; in dealing with life we had to anticipate the beginnings of the future development of Mind. So far from there being a disadvantage in this overflow of these concepts into each other's domain, a truer picture of reality results from such a treatment, which softens the contours of the somewhat too hard and artificial distinctions popularly drawn between them and helps to disclose the underlying unity which pervades them all. It is, however, advisable now to look at the factor of Mind more closely, to define its characters, and to study its functions as an organ and expression of Holism.

It will be readily recognised that the problem of Mind is not for us the same as it is for the psychologist. Psychology treats of the mind in man and the higher animals as a distinct phenomenon by itself, which it analyses and explores in its various elements, and which it studies as a separate department or rather compartment in the total domain of science. For the psychologist the question of boundaries is, therefore, essential; he must demarcate his area of Mind from other areas in the total world of knowledge. He must at all costs vindicate the claims of psychology as a separate science, distinct from the rest. And having with more or less success differentiated the scope of his science from those of other sciences, he then proceeds to explore the details of his science in the manner which is well known to us from the methods and procedure of the great masters of psychology. It is just here in the settlement of boundaries, in the demarcation of the domain of psychology from other domains in science, that the fundamental difficulty for psychology arises. For Mind is much more subtle, pervasive and masterful than life and still more so than matter. Its " field " covers and penetrates the " fields " of matter and life in a way which makes the tracing of hard-and-fast boundaries very difficult, if not practically impossible. It seems to impinge in all directions on areas already apparently securely held by the other departments of natural and biological science; its claims are contested in many directions; and serious doubts arise in how far it really has a territory of its own distinct from other territories in science. The nature of Mind makes this difficulty inherent and irremediable, and psychology as a separate science will always have to remain content with an intensive cultivation of its central area only, and a sharing of the outer marches and outlying territories with the natural and biological sciences. To me it seems that such a condominium of the debatable area, however awkward for psychology, is by no means an unmixed evil for science in general, and that the intimate contact of the different view-points and methods of psychology and the other sciences over this area may prove fruitful and productive of great advances in future. This is, however, remarked by the way. My real point is the difference in the treatment of mind from the standpoints of psychology and of Holism respectively. For psychology Mind is a distinct phenomenon to be studied by itself. For Holism Mind is but a phase, though a culminating phase, of its universal process. The question of boundaries, so fundamental for the psychologist, does not exist for us. From our point of view that is a mere parochial question; for us Mind is not merely a phenomenon of human and animal psychology. We have to trace the connections of Mind with the earlier phases of matter and life; we have, so to say, to lay bare the foundations of Mind in the order of the universe. Mind as an expression of Holism, Mind as an organ of Holism: that is our problem.

We have already seen that the atom and the cell were the two great departures in the upbuilding of the universe, the two great abiding peaks of achievement in the march of creative Holism, which have in turn become the basis and fundamental units of all existence. We now come to the third, which in the order of the universe is perhaps as great a departure, and from our human point of view even more significant than the other two. In Mind we reach the most significant factor in the universe, the supreme organ which controls all the other structures and mechanisms. Mind is not yet the master, but it is the key in the hands of the master, Personality. It unlocks the door and releases the new-born spirit from the bonds and shackles and dungeons of natural necessity. It is the supreme system of control, and it holds the secret of freedom. Through the opened door, and the mists which still dim the eyes of the emergent spirit, it points to the great vistas of knowledge. Mind is the eye with which the universe beholds itself and knows itself divine. In Mind Nature at last emerges from the deep sleep of its far-off beginnings, becomes awake, aware and conscious, begins to know herself, and consciously, instead of blindly and unconsciously, to reach out towards freedom, towards welfare, and towards the goal of the ultimate Good. Mind is thus the organ of control, of knowledge and of values. No wonder that to the young Socrates it came as a great spiritual revelation when first he learned from Anaxagoras that not matter but mind was the ultimate principle of the universe. It is at any rate worthy to be set by the side of the atom and the cell as among the fundamental advances in creative Holism.

It would be an interesting speculation at this stage to pause and ask, from our knowledge of the previous lines of advance in the atom and the cell, what the next step was to be, or rather in what direction it was to be. In what way precisely does Mind fit into the scheme of the earlier structures and mark another step forward in the great line of holistic advance? An answer to this larger, more speculative question may give us some general clue to the nature of Mind as the next great factor or phase in the evolution of the universe, and may form a fitting introduction to the narrower, more practical question of the functions and activities of Mind in the higher animals and man.

Mind is an advance on what has gone before in two directions. And it is the peculiar interaction between the double lines of advance, the intersection of the two curves of advance, so to say, that produces the uniqueness of Mind as a natural phenomenon. In order to appreciate this we have to grasp the point which has been reached in the preceding sketch of the evolution of Holism.

We have seen that both matter and life are structures, and that the advance of Evolution consists in the emergence of ever more complex and intensive structures, ever more complexly and highly organised wholes. In the structures of matter the number of co-operative elements are fewer and their interactions are simpler, so that it is still possible to some extent to trace elementary effects to their separate causes and sources in the structure itself. Structure is dominant and its functions are calculable as elements of structure. As, however, we proceed from physical to chemical structures the fusion and unification of elements and functions become more marked, and the structures become at the same time more complex. When we come to the structures of life we find not only the structural elements far more numerous and the structures far more complex, but also the organisation much more intensive and unified and the functions much more single and unified, individual and unanalysable. In a tree or an animal, for instance, we find an infinity of cells and cell-structures of all degrees of specialisation mutually adapted to and co-operating with each other for the maintenance of a single individual whole in a most wonderful way. We do not ascribe this co-operation and unity of action to some presiding intelligence in the tree or animal. In organism as such there is no psychic control; and yet there is a control so simple, so automatic, so effective as to baffle our powers of understanding. The inner co-ordination and self-regulation in organisms which is the organic phase of Holism is indeed something marvellous, almost something miraculous. And in its way and on its own plane nothing more wonderful or perfect has been reached in the evolution of the universe. Mind with its uncertainties, its aberrations, its failures, seems a mere bungling experiment compared with this massive certainty and regularity. The irregularities and eccentricities of Mind in man compare very unfavourably with the unerring precision and regularity of organic activity and functioning in all highly developed plants and animals. Think of the well-ordered society which constitutes a big animal or tree! Compare the love-making and union and reproduction in plants and organisms with the love-making and union of hearts of humans! Compare the social organisation of insects with our social disorganisation and anarchy, our painful and uncertain social experiments and expedients even in the most highly developed human societies! No, organism has nothing to learn from Mind in the way of regulation, co-ordination or inner control of structures and functions. The self-balance of processes and activities in organism surpasses anything our ingenuity can understand or encompass. It is by reflections such as these that the impression is borne in upon us that Mind is no mere continuation and development of the organic process, but largely a fresh experiment in the universe, an experiment still in the making, and by no means in every respect a successful one. Mind, in fact, is a new structure still in process of making, and not a direct continuation or expansion of what has gone before. It is a superstructure on the basis of the pre-existing physical and physiological structures, and it carries on the task of Evolution on somewhat new lines of its own, and initiated by itself. It has not appeared suddenly and from the blue at any particular point, though its advance may have partaken of the character of a mutation, or a series of mutations. Its primordial roots probably lay in the beginnings of life itself, and in the favouring bosom of life its embryonic structure developed until in time it could appear as an independent factor, with a steadily growing power over life itself. But during all that immense formative period it was but a nursling of life and in no intelligible sense was it responsible for the delicate, complicated, internal self-ordering of the life-structures which must be attributed to another prior factor, or rather to a prior development of the same underlying holistic process of which Mind is a later development.

Let me now turn to the consideration of the double lines of advance along which Mind emerges and pushes forward in its evolution. In the discussion of cell-structures in Chapter IV we noted a double process in Holism, one of which is the regulative universalising process of structural order to which so much attention has been paid in this study, and the second of which I called individuation. Let me here say a few words about the latter aspect of the holistic advance, which remained of a somewhat subordinate character and comparatively minor importance until the appearance of Mind. Holism, as its very idea implies, is a tendency towards unity, a blending and ordering of multiple elements into new unities. From the more or less homogeneous to the heterogeneous; from heterogeneous multiplicity again to greater, more advanced harmony, to a harmonious co-operative ordered structural unity; such a formula may serve as a rough-and-ready description of the holistic process. Thus, for instance, in the process of Evolution we see the advance from material systems to individual organisms. One organism is not merely a duplicate of another, as one molecule of water is a duplicate of another. It is single and individual, with a character of its own. And the element of separate individuality increases as the differentiation and variation increase with the advance of Evolution. Such individual differences tend to increase, and at the same time their blending in the individual tends to become ever more unique. This tendency towards individuation is inherent in the holistic process and receives an immense impetus when the human level of development is reached. Here for the first time individuality acquires its true and full meaning. Everyone knows what is meant by individuality as applied to humans.

Not only are no two human beings alike; their separate characteristic individualities are what is most distinctive of them and what they are known by and what principally determines their relations in life. There is in each human being not only a peculiar blending of characters but also a sense of the uniqueness of this blending, a sense of separate and specific selfhood which constitutes his or her very essence. Humans are not mere units (as material bodies), they are also individuals; they are not merely individuals (like organisms), but also unique selves. Thus is the fundamental principle of individuation finally consummated in the human. The human being is a conscious self, and this selfhood becomes in turn the basis of his Personality, which is the supreme structure yet reached in Evolution and with which we shall deal in the next chapter. It is a striking fact that in the holistic advance as I have sketched it in previous chapters the dominant note and feature of progress is order, with an ever-increasing measure of regulation and co-ordination and control so as to make that order effective; while the feature of individuation is comparatively insignificant. As old as structural order itself in the evolution of the universe, and an inseparable accompaniment of it at all stages, individuation as an evolutionary variation remains in the background, so to say, until the emergence of Mind leads to a rapid and indeed phenomenal outgrowth of this hitherto minor feature. The appearance of Mind, therefore, especially at the human level where it is most marked, seems to constitute a break in the even and regular advance of Evolution, and to mark a new departure of a very far-reaching character. The fact is that in and with Mind a significant change takes place in the relative importance of the two fundamental aspects in Holism. While the aspect of order and regulation continues to develop and grow, the other aspect of individuation pushes relatively much more to the front, and in the latest human phase of evolution not only assumes a dominant importance in itself, but also begins to exert a far-reaching influence on the other feature of order and regulation. That it will and indeed must have such an influence is at once intelligible from the fact that at bottom individuation and regulation are, as we have seen, dual aspects of the same inner process or activity, and any accentuation of the "individuation" factor must at once react on the other "regulation" factor. Thus it is that while in Nature order is of a mechanical character and in the world of life is of an automatic character, certain, regular and unfailing; in man, where the mental factor has come into its own, it is neither mechanical nor automatic, but of a new plastic variable type which we call conscious and voluntary. In fact the whole system of regulation and control is fundamentally transformed; new mental agencies seem to be at work and new categories of description and explanation become necessary. The appearance of Mind has meant, not only an epoch-making development of the feature of individuality, but also and in consequence a new system of regulation and control, not so regular and automatic and effective as the older inorganic and organic systems, and still comparatively vacillating, irregular and uncertain in its action, but vastly more comprehensive and with a power which promises ultimately to give a complete command over the conditions of matter and life. There is a mastery in the new system of control such as was not dreamt of at earlier stages of Evolution. Organic regulation, however vast, elaborate and effective, has nothing of the sheer mastery and domination and free creative power which characterises the new control. Conscious planning on the mental level entirely revolutionises the situation and substitutes freedom and action for the fixed automatic behaviour and routine of the biological order.

These general remarks will serve to place Mind in the history of Evolution, and to show the nature of its relations to what has gone before and what is to follow. It marks a new departure not only in the feature of holistic regulation and control, to which so much attention has been paid in the foregoing chapters, but also and far more specially in the feature of individuation, which up to now has been of an insignificant character, but which from now on begins to assume a dominant position, and to give a new direction and character to the pre-existing system of organic regulation. Mind is not so much a direct continuation of the holistic advance on the previous lines of life as a fresh start, with a new factor pushed to the fore in the process, and a new orientation given to the whole movement. It marks the new stage of intensive individuation which becomes Personality; and at the same time it marks the new system of control which culminates in conscious rational Purpose as a function of Personality. Mind underlies and supports both these great closely related departures in the process of Evolution. Having thus indicated the general function and activity of Mind in the history of Evolution, let us now proceed to look more closely at its nature and character.

Mind has its earliest beginnings in the inorganic structures of Nature already. Disturbance of the equilibrium of physical structures leads, as we have seen, to a state of tension, and a tendency to compensation; and one phase of this tension and compensating movement is seen in the selective action which matter already exercises, and which, as explained in Chapter VII, becomes far more accentuated in the subsequent structures of life. This tension with its selective compensation is without a doubt the original stimulus and source of Mind as well as of life, but the evolution of life proceeded far more rapidly, and completely outstripped Mind in the race which followed. Mind as a matter of fact needed the support of life for its full fruition, and was therefore dependent on the prior development of life. In the course of the subsequent developments this tension underwent two radical changes which had far-reaching effects, as they led directly to the evolution of Mind. In the first place, the tension in the life-structures or living bodies developed (in some unknown manner) an additional intensity which took the form of a vague sense of irritation or discomfort which began to accompany it. In other words, the tension or strain in the living bodies led to the development of this vague sense of uneasiness or discomfort, which had the effect of strongly reinforcing and stimulating the efforts made for the removal of the strain. The successful effort, again, was accompanied by a sense of ease or comfort which must have been a real helpful stimulus and have had considerable survival value for the organisms that developed it. We see thus that the tension or strain came to be accompanied by vague feelings which radically transformed it and gave a different meaning and value to it. The feeling became a potent force working behind and inside the organic system for the removal of the tension or strain. This feeling or sense of comfort or discomfort must originally have been of the vaguest possible character, but at any rate it was a beginning, and it performed a useful function in reinforcing the effort or rather tendency for the removal of the strain or uneasiness. It marked an enormous step in advance, and is probably still exemplified in the "tropisms" which characterise the movements of the most primitive plant and animal organisms.

In the second place, the tension became intensified in another direction. Instead of remaining merely a passive result of the state of disequilibrium, it became an active state or relation between the structure or body affected and the cause of the affection or discomfort. The passive tension became an active *ad-tension* or attention, and in this transformation we reach the most primitive, most characteristic function of Mind. The living organism no longer suffers passively, blindly and in darkness, so to speak. The worm turns upon the source of its torture. The organism begins to *attend* to the source of its discomfort; and this attention, at first vague and diffused, gradually develops, until it becomes an awareness or low form of awareness of its object, and consciousness in its most elementary form thus appears. In its most primitive form it showed itself at quite an early stage in animal development, probably not long after sensori-motor mechanisms had been evolved.

IX MIND AS AN ORGAN OF WHOLES

These are the principal steps in the beginnings of Mind; and whatever immemorial periods this evolution may have taken, and whatever other intermediate phases it may have passed through, in the result the basis of Mind was well and truly laid in the rise of the power of attention, accompanied and stimulated by feelings of comfort or discomfort, and by a certain awareness or consciousness of the object to which attention was directed. As we saw in Chapter VI, it is one of the special effects of Holism to transform passivity into activity, and nowhere has that transformation had a more far-reaching character than in regard to the origin of attention as an active response on the part of organism from the passive state of tension which had preceded it. In this transformation we see not only the origin of mental activity but also a new departure in the system of power, of freedom and of control over its surroundings with which Mind is specially associated.

The actual steps in the evolution of Mind, in so far as they can be traced from available evidence, need not be discussed here in detail. No doubt we have to start with sporadic and uncertain variations in mere organic structural functioning; as these become regular and stereotyped they assume the form, first of tropisms in plants and animals, and then of reflexes in the activity of special organs or cells. Then in the case of animals trains of reflexes are gradually co-ordinated into regular modes of activity of the organism as a whole, called instincts. Sensori-motor co-ordinations are effected, by which the passive influences and effects coming from the outside world are transformed into definite modes of active response by the organism. This active power of response enables the organism to strive more effectively for the satisfaction of its needs, endows it with a definite conative power, so to say. It begins to strive, to seek, to experiment and explore. The original reaction of inorganic, and then of organic, selectiveness has become a real function and capacity of conation. The originally vague and diffused feeling increases in volume and intensity and propels this striving or conation all the more effectively. The awareness or consciousness of objects becomes clearer; consciousness becomes a real illumination of outside objects which before were dark and unknown. It becomes the torch with which the organism explores its way in a dark and somewhat alien world. Consciousness thus increases the influence of the environment on the organism; and its correlative attention *pari passu* increases the power of response and the return influence which the organism can exercise over the environment. This mental activity continues to grow in its double inner and outer aspects, its inner capacity of attention and active reaction, and its outward-facing capacity of assimilating external materials in the form of awareness or consciousness of objects. A metabolism of a higher order than that seen on the biological plane sets in; a new psychological structure has been evolved; and Mind starts on its active and creative career. No elaboration of the steps sketched here can be attempted, and the details must be studied in works dealing with Biology and with animal and human Psychology. We are concerned with the underlying processes and the general character of their results. Details must be left to the special sciences.

I have traced Mind to its dual source, and wish now to draw attention to the consequent duality of Mind itself. Holism on the advanced psychic plane discloses two distinct though interdependent tendencies—the one individual, the other universal—and Mind shows both these contrasted characters, and faces in both apparently opposed directions. Psychologically the duality of Mind is best expressed in the Subject–Object relation which is fundamental for Mind. Consciousness as it develops splits up the indefinite mass of experience into two definite aspects: the self or Subject, which is conscious or attending, the Object, which it attends to or is conscious of. "The Subject—conscious of—an Object" is thus a general formula for all experience of a mental character. The Subject is not before the Object, nor the Object before the Subject, but both arise simultaneously and *pari passu* in the mental activity which we call consciousness. The world is not the creature and result of the Mind, as idealists would have it; nor is the Mind the resultant of external stimuli on the brain, as the materialists would have it. Experience is one; and experience as it becomes conscious differentiates or unfolds itself into the Subject–Object relation. They are the double aspects of experience at its conscious level, and reflect but the duality of the source of Mind itself. The inmost nature and essence of Mind is this activity which appears as consciousness and the Subject and Object aspects which crystallise out of it in experience. They are at bottom and in real truth not independents, but dependent correlates in the psychic medium called consciousness. A clear and firm realisation of this fundamental fact is basic for all true science and philosophy alike. We saw in Chapter VII what insoluble problems arise for both science and thought from hypostatising Mind and Body as independent reals or substances. Here we are at the tap-root of this source of error. Mind-and-Body is but a particular form of the general Subject–Object situation. They are not independents, they are interdependents; they are poles in the field of Mind; they are elements or rather aspects of the same reality given in solution in experience and precipitated from it by consciousness—to use a chemical metaphor. Out of this fundamental unity Mind in the larger sense has elaborated our experience of both the inner and the outer worlds, of the self and the external universe. It is the business of psychology to show how this has been done, and to trace the progressive stages in this constructive process. For me it is only necessary here to emphasise that no correct interpretation of experience is possible unless we bear in mind that both the Subject and Object aspects are absolutely essential to it. Subject and Object are held together in experience as necessary elements in the unity of Holism from which both are differentiated. Neither element can be ignored in our reading and exploration of experience. The Einstein standpoint of Relativity is not only the soundest science, it is fundamental to psychology. The world in experience is at bottom *my* reading of the world in which *I* am the centre reference, where the system of co-ordinates of measurement is *my* private system; and the space, time and experience which go to the making of it are *my* space, time and experience. Objectivity and universality are indeed attainable, but only from a subjective and individual starting-point and centre of reference. Individuation is bound up with reality on the psychic plane. What has been looked upon as a reproach to psychology, namely, the essentially subjective standpoint which rules it, the self which it discloses as central to all our experience of reality now appears to be equally necessary for all the other sciences as well. Natural Science, which has always prided itself on its objectivity and freedom from all subjective consideration, now finds that after all psychology has been right; that the despised subject or Self or psychology not only is a real factor in the universe, but is central to all true knowledge of it. Here at last natural science and psychology can clasp hands and together in true partnership go forward on the great adventure of knowledge which is their common task. Their separation has been a calamity to both; their reunion will prove fruitful beyond our fondest dreams. Einstein's great achievement is but the first-fruits of that reunion.

Let us for a moment look a little more closely at this individualistic aspect of Holism in its higher developments. We have seen how it begins as physical and chemical affinities and selectivities; later on in the region of life we saw it appearing as organic selection and appetitiveness. On the mental level again it emerges as a certain striving, a seeking, a conativeness, which, when attention has risen to the level of consciousness, becomes purposive. Thereafter the individual no longer floats forward on inorganic or organic drifts, tendencies and appetites, but begins to direct his course according to conscious voluntary purposes. The individual makes his own plans and no longer automatically follows Nature's plan. Conscious ends emerge; things the individual strives after as desirable and good attract him more than the unconscious organic urge behind, the *vis a tergo,* propels him; the pulls in front begin to dominate the pushes behind. The desired things become the Values, which intelligence illuminates and magnifies and emotions suffuse and intensify until they become the dominant Ideals of action. We see the rise not only of new mental activities but of new categories such as Purpose and Value, which were not possible or necessary on the organic level. It is evident that these new activities, such as attention, consciousness, intelligence, emotion and will, as well as the new categories which accompany them, are all in line with the individuational development of Holism, and mark so to say a deviation from the direct line of organic regulation and systematic co-ordination which characterised Holism in its earlier organic development. They make directly for the development of the individual, of the self, of the Personality. Holism seems for the moment to depart from its vast plan of extensive co-ordination and harmonisation in order to foster little centres of intensive Wholeness in individuals, to place the little wholes before the great Whole, and to abandon universality for individuality. In fact the largely individualistic nature of Mind makes it apparently a deviation from the universal order. Mind appears as a rebel in the universe, whose self-centredness and purposeful striving might and largely does make for disharmony and disorder rather than for peace, order and harmony. Thus the great Ethical problem arises; thus the conflict between individual ends and purposes on the one hand and universal claims and rights on the other comes to the surface on the psychic plane of Evolution. Mind the rebel has appeared, Self the anarchist has emerged, and the ancient order of the universe is profoundly disturbed. The new psychic individualistic situation applies a most searching test to the foundations of the holistic universe. The war in heaven has broken out, the archangels have revolted. Who will win, and what is the character of the new peace going to be? Such is the question which we now proceed to consider.

The first and most important point to make is that we should not be misled by our own metaphors. Individuation is but an aspect and no more, is but one aspect of the holistic advance. The other, universal, aspect remains of fundamental importance, although for the moment it may be and probably is pushed somewhat into the background. The two aspects are complementary and interdependent, and each has a vital grip on the other. The individual is going to be universalised, the universal is going to be individualised, and thus from both directions the whole is going to be enriched. The individual development is necessary for the advance. Organic regulation, however great an advance on inorganic structural order, is not enough. It is still too mechanical and rigid; it is still too external in character. It must acquire new characters of internality; there must be more self-regulation and less external regulation. There must be more inner intensive mobility, more plasticity and less rigid regulation. Plasticity, freedom, creativeness are necessary for the new groupings and structures which are to arise on the psychic level. The higher metabolism of mind demands more freedom from the routine of organic regulation. The area and degrees of freedom must be indefinitely enlarged, and in the Self a new organiser of victory must be constituted. That is the one side of the dual advance; but the other is there all the time and in the end just as important. Thus there must also be a higher order, a more developed and enriched universality, a more coherent objectivity. More wholeness not only means a deeper, more intensive individuality in the Self, but also a more perfect order in the structure of Reality. And the two must interpenetrate each other and mutually transform each other in that unity which is the whole. And that is exactly what happens in the evolution of mind in man and animals, and in the development of mind in the human individual. The selfish appetite is gradually curbed and subdued and co-ordinated with other motives; the

IX MIND AS AN ORGAN OF WHOLES

conative activity is not merely self-regarding but gradually becomes linked with the interests of others, and finally becomes an impersonal endeavour towards the Good. A new era of adjustments and co-ordinations sets in, and the individual on the psychic plane pursues the double task of self-protection and perfection of the All. And the two mutually and reciprocally influence and modify each other and shape an ideal of Good which incorporates elements from both. Holism has narrowed itself into the individual only thereby to advance to a more perfect all-embracing order. The apparent retreat to the individual level is merely for the purpose of a greater advance towards wholeness. The newer, deeper Self becomes the centre for a fresh ordering and harmony of the universal.

The possibility for this great transformation is given in the very nature of Mind; for Mind is not merely conative and purposive. It is also rational, it is the basis of the Reason. And Reason becomes the basis of the new order in the universe. It is not only the principle of order in the Self, but also the link which binds the Self and the Not-self into a whole. Reason is the organ of universality, of the deeper, more intensive universality of the spirit. Reason is largely creative of the new structures of Reality and Truth. In the Reason, Mind, instead of pursuing its individualistic, purposive activity, resumes the primeval march of Holism towards more regulation, a higher co-ordination and a greater order. Our will is the urge towards self-expression, and is therefore the organ of individuality. "Our wills are ours to make them Thine." In other words, our will is individualistic and has to be harmonised and through effort and struggle to be adjusted to higher ethical and spiritual ends and ideals. But our Reason is in its very essence more than individual; it is expressive of universality; it is a part of that Order which regulates the universe, and in a deep sense it is a creative factor or co-creator of that Order. Through our Reason we partake of universality and are members of the everlasting Order of the universal.

Mind in its rational, as distinguished from its purely

conative activity, is in the direct line of Evolution from organic regulation: psychic reason is the direct descendant of organic regulation. This would appear from the simple fact that the central nervous system is the physiological organ of regulation and co-ordination in the living body; and the brain, which is nothing but the crowning development of the nervous system, is again the organ and physiological correlate of Mind. In other words, as the brain is merely a development of the nervous system, so the Mind is nothing but a development of prior organic regulation. Mind *qua* Reason is thus the organising principle, the principle of central control and co-ordination, and carries on the tradition and evolution of Holism in the direct line, so to say.

And in the entire range of its rational activity Mind shows the same synthetic co-ordinating character. This could easily be proved by running through the psychological functions, ranging from their beginnings in attention and passing through sensation, perception, imagination, conception, and on to judgment or reasoning. In every case an ordering synthetic activity is at work producing structural masses of experience arranged on definite ascertainable principles of selection and grouping. These products of the rational activity of Mind are not mere artificial aggregates, mere assemblages of psychological items arranged according to mechanical principles or so-called laws of association or rules of logic. On the contrary, they are synthetic unities of an advanced holistic character. A percept or an image or a concept is a holistic unity, built up out of a mass of materials, on definite principles of cohesion and co-ordination. So too is a judgment. It is the business of psychology to study these syntheses and their structures, and the principles according to which they are formed, and connected with each other. Mind in its rational activity is thus synthetic and co-ordinative through and through, and its products are synthetic, organic and holistic in a marked degree.

Mind the organiser transforms, reorganises and reconstitutes even the individualist Self. The rebel in the end has to submit and swear fealty to the controlling power. Indeed the purely individualist Self or mere individual is a figment of abstraction. For the Self only comes to realisation and consciousness of itself, not alone and in individual isolation and separateness, but in society, among other selves with whom it interacts in social intercourse. I would never come to know myself and be conscious of my separate individual identity were it not that I become aware of others like me: consciousness of other selves is necessary for consciousness of self or self-consciousness. The individual has therefore a social origin in experience. Nay, more, it is through the use of the purely social instrument of language that I rise above the mere immediacy of experience and immersion in the current of my experience. Language gives names to the items of my experience, and thus through language they are first isolated and abstracted from the continuous body of my experience. Through the naming power of language, again, several items of experience can be grouped together under one name, which becomes distinctive of their general resemblances, in disregard of their minor differences. In other words, the power of forming general concepts becomes possible only through the social instrument of language. Thus the entire developed apparatus of thought with which I measure the universe and garner an untold wealth of personal experience is not my individual equipment and possession, but a socially developed instrument which I share with the rest of my fellows. Nay, my very self, so uniquely individual in appearance, is, as I have said, largely a social construction, and rounded out of the social intercourse and psychical interaction with my fellows. The individual Self or Personality rests not on its individual foundations but on the whole universe. Psychology conclusively proves that, and Holism but accentuates it by tracing the individual to his sources in the whole. The individual Self is not singular, springing from one root, so to say. It combines an infinity of elements growing out of the individual endowment and experience on the one hand and the social tradition and experience on the other. All these elements are fused and metabolised into a holistic unity which becomes a unique centre in the universe and, in a real sense, of the universe. Nowhere in the world do we find a greater intensity of the holistic effect produced than in the individual Self or Personality. And yet even there it is by no means complete, for the individual Personality, as we shall see, still shows a discordance of elements which leads to most of the great problems of thought and conduct. The point I am trying to make, however, is that the apparently individualist Mind is in reality deeply and vitally influenced by the universal Mind; and that the individual self only comes to its own through the rational and social self which relates it organically to the rest of the universe. It is rooted in and dependent on the greater whole, and only to a minor extent a rebel against its controlling influence. The immense power of Mind is shown in the way in which, out of the simple data of the rudimentary Self and its experiences, it has raised the noble superstructure of the human Personality. Mind here appears as the great creative artist. But it is more than that; for its work is no mere picture of reality, but is reality itself. It is the great archetype of the artist, and it has this pure creative power because it is but a form, a phase of the supreme creative activity in the universe.

When Mind comes to apply its conceptual system to its experience of the world, we see the same synthetic holistic activity at work. At first crude, naïve experience is simply taken at its face value, and from it a rough empirical order is constructed which is sufficiently correct for all ordinary purposes, and may fairly be called the world of common sense. Of course even this common-sense empirical order will vary widely at different levels of culture. But in every case it is a first rough approximation and a grouping, ordering and arranging of experience according to the standards and the needs of the common man at that level of mental culture. It is a more or less faithful reading of ordinary experience, and although it contains many discrepancies and contradictions, it is on the whole a more or less connected, coherent *Weltanschauung* with a fair correspondence to the facts of direct observation.

From this common-sense world of experience is gradually evolved a more correct and refined system of experience. Anomalies are gradually eliminated and the logical rational character of the system increases. And this refinement continues until in the scientific conceptual system the empirical common-sense order is completely overhauled and reconstructed. In this system scientific concepts and entities corresponding to nothing in empirical experience become of fundamental importance, because without them not only empirical experience but also the more refined observations of Science become utterly unintelligible. The matters dealt with in most of the preceding chapters belong to the scientific conceptual system, as they base the world of experience on real or hypothetical entities and factors which, although they lie beyond the world of direct experience, are yet necessary for the rational order and coherence and comprehension of the facts which do fall within the range of ordinary and refined scientific experience. In this system not only is the seen order made to depend on an unseen order of ideas of extraordinary refinement, but the immense movements and changes of the universe are referred to a very small number of fundamental principles which appear to govern all happening in the universe. From this esoteric system to which Science is more and more tending have been removed most of the anomalies, incoherences and discrepancies of the empirical order. Many final difficulties still remain, some of which may perhaps never be eliminated. But on the whole the system of Science is rapidly becoming a great rational body of experience and thought, closely articulated in all its details, and held together by simple principles of the widest sweep. The system as a whole represents the proudest achievement of the Mind in its rational activity as the regulative co-ordinating principle in the universe. Mind as the principle of the rational construction of the universe here reaches its highest expression. Professor L. T. Hobhouse in his great work on *Development and Purpose* has with a master's hand traced the development and interpretation of experience from its humble naïve beginnings to its culmination in the vast conceptual system of Science. I must here rest content with the preceding summary statement. What I have said will, however, I hope, suffice to show that Mind the Rebel is only one aspect of holistic activity; and that Mind the Organiser, Mind the Central Control in our experience of the world is the other equally true complementary aspect. Behind both aspects is that inner creative Holism which has flowered into the human Mind and Personality on the one hand and into the grandeur of form and content of the infinite universe on the other. The theory of Holism thus carries the scientific system of experience another step further, and tries to read in the riddles of Science still deeper and more ultimate concepts of reality.

Mind has been here described as a new variation or mutation or series of mutations in holistic Evolution, in some respects antithetic to and at variance with its main trend. But its final result is immensely to enrich the main process. Mind has made all the difference to the later and latest stages of Evolution. Without Mind the organic and regulative process of the universe, vast and magnificent in any case, would have been at best but a tame affair. The universe would have moved forward, as it were in a dream, with an unearthly regularity and majesty of movement. Its process would have become ever more complicated and ever more frictionless, as of some sublime animated machine, great beyond all power of conception. All elements of discord and disharmony would have passed away from its vast cosmic routine. But it would have gone on sublimely unconscious of itself. It would have had no soul or souls; it would have harboured no passionate exaltations; no poignant regrets or bitter sorrows would have disturbed its profound peace. For it neither the great lights nor the deep shadows. Truth, Beauty and Goodness would have been there, but unknown, unseen, unloved. They would have been cold and passionless like the distant stars, and would never have become the great ideals thrilling and inspiring men and women to deathless action. Love would have been there, but not the immortal emotion which mortals call by that name. Into that great dream-garden of Eden, Mind the Disturber has entered, and with Mind sin and sorrow, faith and love, the great vision of knowledge, and the conscious effort to master all hampering conditions and to work out the great redemption. To the music of the universe there has thus been added a new note, as of laughter and tears, a new undertone of the human, which transforms and enriches all the rest. It is no longer a song of the Golden Reign of the Elder Gods, but of the intertwining of the Cosmos with human Destiny, of the suffering which has become consecrated and illuminated by the great visions, of the magic power of knowledge to work out new enchantments to break the dumb routine, to set the captive spirit free, and to blaze new paths to the immortal Goal. Mind has thus added an infinity of light and shade and colour, of inward character and conscious content to the great process in and from which it has emerged. Without Mind the universe would have been an altogether dull affair, however unimaginably grand in other respects. Even its aberrations have been woven into the new harmonies; its eye has beheld the greater lights, and knowledge has given it the key of power and mastery over the conditions which previously towered like an unscalable mountain escarpment athwart its path of progress.

Let us dwell for a moment on this new power and mastery which Mind has brought on the scene. Knowledge is power, and it is unnecessary for us here to trace in detail the steps by which the present power and mastery of Science over material conditions have been acquired. Life below the mental level strengthens the innate capacity to react to external influences of a harmful or beneficial nature by various movements which lead successively to the tropisms, reflexes and automatisms of the lower organisms. When Mind appears as an active factor, this power of regulating movements is greatly enlarged and intensified, until we see the sureness and delicacy of the instinctive reactions which characterise all Mind in its subconscious levels. It is, however, when consciousness appears that an immense accession to this power of control is brought about. Consciousness, as we have seen, is a power of illuminating objects in the field of experience. The organism through this power of illumination can gradually arrive at a fair knowledge of its surroundings in so far as they are harmful or beneficial to it. Its power of selection is thus more surely guided, and it learns to know accurately and easily what to avoid and what to welcome. When the human level is reached a revolution in the conditions of knowledge is effected. The human mind can make its own combinations and correlations from the materials with which it finds itself surrounded. It can, therefore, in a large sense make or mould its own environmental conditions, and thus eliminate or neutralise hostile influences and reinforce favourable conditions. This is already the case on the empirical level of knowledge; it is far more the case where the empirical stage has been passed and the developed scientific stage has been reached. Here the mind does not wait on events, but moulds and creates events through its control of the appropriate conditions. In this way the development of the several sciences has meant continuous increase, not only of knowledge, but of real power over the material and other conditions of life. Here again Mind the Organiser or Correlator has shown its creative power in shaping the conditions which surround its activity. Instead of being the slave of these conditions it gains a more or less complete mastery over them. It can at will bring about those combinations and selections which will assist or further its purpose, and it can, through selective manipulation of the surrounding conditions, neutralise or cancel out any which are unfavourable to the execution of its aims. Knowledge thus becomes an efficient instrument of the will; and where the will itself is nobly trained, guided and controlled, the individual acquires and wields an almost unlimited power for Good. Thus is freedom at last achieved over the dominance of the conditions of life, and Mind assumes the sovereignty to which it had been destined from the beginning as the successor to Life and Matter.

In the exercise of its free and unhampered right of self-determination, Mind on the human level proceeds to create to a large extent the appropriate conditions for its own development. Instead of remaining dependent on the natural environment, Mind builds up a vast social environment for itself. It builds up a far-reaching social structure with institutions of all sorts which are intended to develop and educate the human groups and individuals, intellectually and morally, to facilitate intercourse and co-operation among them, to declare and safeguard their rights, and to protect them against the hostile influences of the animate or inanimate environment and of other groups of humans. Thus language arises as well as the institutions of marriage and the family, of religion, law and government, and all the other numerous forms into which social beliefs and practices are embodied. The very laws of organic Evolution seem to be modified by this great transformation. In the organic sphere we saw the individual adapting itself or being adapted to the environment as the imperative condition of its survival. Here we see the environment being more and more adapted to the individual. The individual appears as the creator, the environment as the creature, the house it makes for its habitation, so to speak. In the organic sphere we saw the individual inheritance and variations incorporated into the individual organic structure and thus preserved for the future. Here we see social traditions take the place of this individual structural heredity. The human individual does not find himself over-burdened with an impossible structure, with a load of inheritance which would be more than he could bear. The load is mainly shifted on to the ampler shoulders of the social tradition. The human individual has the good luck to find himself born into an environment which largely performs the hereditary function, and all that he is called upon to do is to assimilate this environment, and so to obtain command of its gathered resources. Language, customs, writing, literature, history, knowledge and empirical practice are all storehouses of traditional information at the disposal of the human individual who learns their use. Herediy with the human individual comes more and more to mean, not (as in the case of animals) the predisposition or capacity to act or react in certain definite ways, but the general capacity of experience, the capacity to *learn* or acquire in the individual life the power to act in an indefinite number of ways. In the human inheritance general educability takes the place of definite specific hereditary functions. Whereas an animal is born with the ability to perform a certain limited number of functions, the human individual is born with the general capacity of educability or being educated to learn an indefinite number of functions in his lifetime. The animal is still under the domination of his physical structure, and in his action is limited to the functions inherited with this structure, with a very limited range of learning new actions. The human individual, on the contrary, finds himself but little restricted in his development by his hereditary structure, and finds himself blest with an almost unlimited adaptability and capacity for experience and knowledge. In other words, the inheritance of Mind supersedes the organic inheritance more and more. With an animal definite modes of functioning are inherited; with the human individual general mental plasticity is chiefly inherited. And the definite specific modes of functioning which an animal inherits with his physiological structure, the human individual learns and acquires from the social tradition into which he is born. Nothing shows more clearly the revolution which the appearance of Mind has wrought than this far-reaching transformation which it has effected in the methods and procedure of organic Evolution. On the animal plane structure still largely determines function, but on the human plane mental plasticity so dominates everything else in the inheritance that the importance of structure is completely dwarfed, and it appears as a subordinate factor in the total human situation. Even so, however, it retains a great importance which often comes out in dark and unexpected ways in the individual conduct.

The advent of Mind has undoubtedly meant a large correlated development of structure, especially in the human nervous system and brain, which is, of course, far larger and more complex than that of even the highest anthropoid. But even so the rôle of structure is comparatively less prominent in man than it is at earlier phases of organic Evolution, and its functions have not only been fundamentally transformed, as we have just seen, but have also relatively vastly increased in significance. So much so indeed that structure in man becomes of merely secondary importance, while its mental functions become all-important. The superstructure of Mind is immeasurably greater than the brain or neural structure on which it rests, and is something of a quite different order, which marks a revolutionary departure from the organic order whence it originated. Under these circumstances the question of primacy as between the Mind and the brain is deprived of all real importance. It is not a question of origins but of values, to which there can be but one answer. By whatever standard of value it is measured, Mind has risen above its physiological source as high as, or even higher than, life has risen above its inorganic beginnings.

From the question of structure we pass naturally on to consider the "field" of Mind. The field of Mind differs in character from the field of matter or of organism. It is neither physical nor physiological, no more than Mind itself is. Mind is a new type of structure of the immaterial or spiritual kind, and so also is its field. In Mind there is a central illuminated area, the area of full consciousness, which is directly open to inspection and observation. Taking this area as the central structure of Mind, the "field" of Mind then comes to mean that area of its functions and activities which falls below the "threshold" of consciousness, which remains unilluminated and dark, which cannot, therefore, be known by direct inspection and which, as in the cases of the other fields, can only be ascertained by its indirect effects. The field of Mind in this sense has been the subject of much psychological speculation and discussion, and we may therefore here rest content with a very summary statement, which will as far as possible be confined to the holistic aspects of subconscious mental activity.

The activities of Mind below the level of consciousness are most important for Mind as a whole. It is in this subconscious area or field of its activities that Mind especially feels the pressure of the past and to some extent the pull of the future. The time factor is even more important for the field of Mind than it is for the other fields previously considered. The central structure of Mind functions in the full blaze of consciousness in the present; but it is surrounded by a field of the greatest importance where the past and the future respectively hold sway. Mind, therefore, integrates the past and the future with the present; mental activity is a synthesis which unifies all its time in the present moment of functioning. And it is thus enabled to act with far more holistic effect than either matter or life is able to do. Let us consider for a moment the influence which the past and the future exercise on the present in mental activity.

The contribution of the past is twofold. In the first place there is the experience of the past in the individual life which has fallen into the background of the Mind and is no longer directly remembered. Yet this experience, as is well known, has a most powerful influence on the conscious present of the experiencing subject. Even the unremembered past experience is not dead, but alive and active below the level of consciousness. In the debating chamber of the present it may not speak, but it *votes*, and its silent vote is often decisive. Mind does not work in water-tight compartments, its past experience is integral with its present action. Its procedure is entirely massive, integral, holistic. Memory, the great basic bond of individuality, binding together and fusing all the past phases and experience of the individual with the present into one unique whole which is himself, operates below as well as above the level of consciousness; essentially it forgets nothing and leaves behind nothing of the past. Remembered or unremembered, the past exerts its full force on the present experience.

The second contribution of the past comes from farther back. It is the contribution of the hereditary structure as modified by ancestral experience, which lies behind all individual experience. And in many ways its contribution is even more significant than that of past individual experience. It gives us our fundamental bias, our points of view, our temperament, our instinctive reactions and our particular individual ways of looking at persons and things. There is in each human individual a distinctive basis of Personality composed of these elements which cannot be traced to individual experience and which is given by his hereditary structure and ancestral past. In many ways it is the most important part of our personal make-up. It is not conscious or critical or rational in its activity, but it constitutes the permanent background of the Mind and the Personality behind all individual experience and development. Experience, reasoning, criticism usually make no impression on it. I like or dislike somebody instinctively and at first sight, and nothing thereafter alters my attitude to him. There is nothing analytical about it, and its action is purely massive. Generally the result of this massive hereditary memory could be best described as a "feel" or sensing, an intuitive reading or subconscious judgment of a person or thing or situation, which cannot be further analysed to any good purpose. Great wisdom and judgment no less than prejudices and passions usually have their source in that distant past and rest on no analysable evidence in the individual experience. It will thus be seen that this contribution of the hereditary past is also decidedly of a holistic character. The importance given to it by the recent development of Psycho-analysis need only be mentioned here.

What has been said so far will be generally admitted. To me, however, there is something even more decidedly holistic in the hereditary factor. To me the ordinary senses do not exhaust the possibilities of sensuous intuition in the human mind. These senses have been differentiated and evolved out of a common pool; but the pool has not in consequence dried up and ceased to be the ultimate sensuous source of Holism in the human Mind. Is it a far-fetched idea to assume that behind the special senses and their evolution, and *pari passu* with their evolution, the mother sense from which they were evolved has also silently continued to grow and evolve as the binding, uniting, cementing element among the deliverances of the special senses? There is a subtle, profound, synthetic activity at work among our sensations and intuitions which cannot be ascribed to the ordinary conscious activities of the Mind. All the wholes we see in life as persons or things are composed of contributions from all or most of the special senses, so utterly fused with each other that disentanglement becomes practically impossible. And these uniquely unitary wholes exist for us from the early beginnings of sensation and perception. So, for instance, the unique whole of the mother is present to the young baby from the early weeks of its life. Is there not some subtle fusing, unifying sense at work *pari passu* with the several differentiated senses? Is there not a sixth sense, the *sensus communis* from which the others have been derived without exhausting it, and whose development has kept pace with their development? Such a sense would not be particularly noticed as its activity is ordinarily and as a matter of course apt to be ascribed to and apportioned among the other senses. But the coherence of the deliverances of the several senses and their fusion into unitary wholes cannot be ascribed to some assumed attraction for each other on their part! It is the *Mind* which fuses and unites them; and if it is the mind, it must be a sensuous element or factor in the mind over and above these specialised senses. To me it seems a simple and plausible idea that there is in the mind more power of sensation and intuition of the synthetic type than is to be found in or between the special senses. Otherwise I find the unities underlying both the subjects and objects of experience inexplicable. I am not sure that our massive sense of reality, of the reality of the external world, for instance, is not to be traced in a large measure to the influence of this deeper sense behind the other senses.

Psychologists believe in a general sensibility which shows itself not only in the vague internal organic sensations, but also in the other diffused states of our bodies resulting from light or warmth or other physical conditions. They have, however, not given sufficient attention to the subject, which seems to me of great importance in the synthetic make-up of Mind.

The obscure subject of Telepathy seems to me to fall within the "field" of Mind, and possibly to involve a form of sensuous intuition which cannot be attributed to any of the special senses. The phenomena of Telepathy are mostly associated with striking experiences or topics falling within the area of conscious thought or discussion. It is possible, therefore, that the form of sense which becomes active in Telepathy is closer to thought than the ordinary senses, that it lies between thought and these senses, and that its subtle activity reveals the thought where these senses cannot do so. In short, it may be in the nature of what has been called an "intellectual intuition." Just as all sensation involves thought, so all thought involves obscure diffused sensation. And it is possible that in this area of obscure sensation at the bottom or on the fringe of thought the explanation of telepathic sensation may lie. And subjects who have this form of sensitiveness may sense the thought itself or other conscious experience of other individuals without communication in the ordinary ways of the senses. This sensitiveness may be a form of the above-mentioned universal *sensus communis*, the nature and functions of which are of great importance for the mind as a whole and are well worth exploring by psychology. Here I can only incidentally mention the matter.

So much for the contribution of the past to the present conscious activity of Mind. It will be noticed that it is of a highly synthetic or holistic character and that, whether it operates as subconscious sensation or subconscious judgment, it supplies much of the cement which Mind requires for its constructions and syntheses on the conscious level.

The most significant element, however, in the "field" of Mind concerns the future, and makes the future an operative factor in the present mental activity. Mind does this through purpose; purpose is the function of Mind by which it contemplates some future desired end and makes the idea of this end exert its full force in the present. Thus I form a purpose to go on a hunting expedition for my next holiday, and this purpose forms a complex synthesis and sets going a whole series of plans and actions all intended to give effect to the purpose. Thus in purpose the future as an object in my mind becomes operative in the present and sets going and controls a long train of acts leading up to the execution of the purpose. The conscious purpose, the end as deliberately envisaged and intended, falls, of course, within the conscious inner area of Mind; but numerous subsidiary elements in the plan would operate subconsciously and thus affect only the field of Mind.

It will be noticed that purpose or purposive activity involves much more than merely the influence of the future on the present. Purpose is the most complete proof of the freedom and creative power of the mind in respect of its material and other conditions, of its power to create its own conditions and to bring about its own situations for its own free activities. My purposive action is action which I have myself planned, which is not impressed on me or dictated to me by external necessity, and for the performance of which I take my own self-chosen measures. Through purpose the mind becomes at last master in its own house, with the power to carry out its own wishes and shape its own course, uninfluenced by the conditions of the environment.

Again, purposive activity is peculiarly holistic. Elements both of the actual past and of anticipated future experience are fused with the present experience into one individual act, which as a conscious object of the mind dominates the entire situation within the purview of the purpose or plan. It involves not only sensations and perceptions, but also concepts of a complex character, feelings and desires in respect of the end desired, and volitions in respect of the act intended; and all these elements are fused and blended into one unique purpose, which is then put into action or execution. Purpose is thus probably the highest, most complex manifestation of the free, creative, holistic activity of Mind. Purpose is the door through which Mind finally escapes from the house of bondage and enters the free realm of its own sovereignty. The purposive teleological order is the domain of the free creative spirit, in which the ethical, spiritual, ideal nature of Mind has free scope for expansion and development. The realm of Ends, as Kant has called it, the realm of the great Values and Ideals is the destined home of Mind. And Holism it is that has guided the faltering footsteps of Mind from its early organic responses and strivings and automatisms through the most amazing adventures and developments until at last it enters into its own.

Let me conclude with a few further remarks on the holistic aspect of mental activity.

In Chapters VI, VII and VIII I endeavoured to show, I trust not quite in vain, how Holism as a concept and an active factor can be made fruitful in biology, and can give us a standpoint and method of dealing with the problems of life which very much facilitates the proper solutions. The problems of Mechanism, of Body and Mind, of Evolution, and many others, all wear a different and more tractable form when viewed from the standpoint of Holism. I think I may fairly claim that the concept and function of Holism will prove even more valuable in the study of Mind, its activities and problems. Mind as a higher, more evolved organ of Holism will naturally exemplify the holistic standpoint more fully than life does. And the holistic conception of mental activity appears to me to be specially helpful and to steer clear of most of the errors and misconceptions which have beset that difficult subject. The holistic conception of mental functioning explains at once why all the psychological activities—from attention to judgment, and not only in intelligence but also in volition, action and emotion—are synthetic in character, and result in associations, syntheses, groups, things, bodies and wholes. It was the unique service of Kant to psychology to discover the presence of the synthetic judgment at work already in the earliest forms of sensation and intuition; and he signalised his discovery by applying to the subject in experience the truly Olympian name of the "synthetic unity of apperception." The view of Mind as Holism leads straight to the same result, and quite simply and without the necessity to resort to any cumbrous psychological or metaphysical apparatus. The activity of Mind at all stages and in all forms is holistic, structural and synthetic, and its products show the same characters. The discriminative, selective, ordering, synthetic character which mental activity shows in all its higher operations is already fully present in its early beginnings, and flows indeed from its very nature as Holism. The various mental functions as dealt with in psychology, therefore, are simply so many examples of holistic activity on the mental plane. Analysis and discrimination may appear to be unholistic, but even they are but means to an end in the synthetic process; the analysed and discriminated elements being but a stepping-stone to more effective selective syntheses and groupings. It would be both interesting and useful to run through the various psychological activities and to show how they all exemplify and indeed flow from the nature and concept of Holism. The task would be easy, but it must be left to others. In this sketch of the subject of Holism I can but confine myself to tracing the larger outlines, and leave particular clues to be followed up by others who may feel interested in the subject.

Equally fruitful, in my opinion, will the application of Holism prove to the problems of metaphysics and the other higher disciplines of Mind. There is not a problem of Metaphysics, of Ethics, of Art and even of Religion which will not benefit enormously from contact with the concept of Holism. Indeed the concept and standpoint of Holism may transform many of their fundamental concepts and render obsolete much of the somewhat barren analytical speculations which are still current in philosophy. In this place I must be content with this reference to the possibilities which lie along the path of holistic argument and research. These applications of the concept of Holism lie beyond the scope of the present work.

CHAPTER X

PERSONALITY AS A WHOLE

Summary.—Personality is the latest and supreme whole which has arisen in the holistic series of Evolution. It is a new structure built on the prior structures of matter, life and mind. The tendency has been to look upon it as a unique and isolated phenomenon, without any genetic relations with the rest of the universe. Our treatment, however, shows it to be one of a series, to be the culminating phase of the great holistic movement in the universe.

Mind is its most important and conspicuous constituent. But the body is also very important and gives the intimate flavour of humanity to Personality. The view which degrades the body as unworthy of the Soul or Spirit is unnatural and owes its origin to morbid religious sentiments. Science has come to the rescue of the body and thereby rendered magnificent service to human welfare. The ideal Personality only arises where Mind irradiates Body and Body nourishes Mind, and the two are one in their mutual transfiguration.

The difficult question of the Body-and-Mind relation, already referred to in Chapter VII, arises once more in connection with Personality. As there pointed out, the root of the difficulty lies in the separation of the elements of Body and Mind and their hypostasis into independent entities. They are not independent reals; disembodied Mind and disminded Body are both impossible concepts, as either has meaning and function only in relation to the other. The popular view of their relation as one of mutual "interaction" is not correct, as Mind does not so much act on Body as penetrate it, and thus act through or inside it. "Peraction" or "intro-action" would be preferable to "interaction" as a description of the relation of Mind to Body. The extreme difficulty of conceiving how two such disparate entities as Mind and Body can influence each other has led to various theories of their interrelation, such as—that God is the medium and agent between them (Berkeley and Geulincx); that their separate action is inwardly brought into accord by a Pre-established Harmony (Leibniz); that they are but two modes of action of the one underlying Substance (Spinoza). The fact is that all these theories have an element of truth; the real explanation being that Mind and Body are elements in the whole of Personality; and that this whole is an inner creative, recreative and transformative activity, which accounts for all that happens in Personality as between its component elements. No explanation will hold water which ignores the most important factor of all in the situation, and that is the holistic Personality itself. Holism is the real creative agent, and not the entities suggested by the above philosophers.

We see this same creative Holism in Personality when we come to consider our inheritance from our parents and ancestors, which consists of a definite animal body slightly differing from theirs, and a mental structure somewhat resembling theirs. My Personality itself, however, is indisputably mine, and is not inherited from them. It may in some respects resemble theirs but its very essence is its unique individuality. The fact here too is that Personality is a unique creative novelty in every human being, and that no explanation which ignores this creative Holism can even pretend to account for Personality.

For psychology and epistomology the individual Subject is the centre of orientation in all experience and reality; it is the Subject of Experience to which all the rest is the Object of Experience. The appearance of Personality, therefore, marks a new and fundamental departure in the evolution of the universe. These disciplines concentrate on the Subject as the centre of reference for experience without, however, paying sufficient attention to the nature of Personality in other respects. Ignoring the individual uniqueness of the Personality in each case, psychology deals with the average or generalised individual; and then only from the purely mental point of view, which is but one aspect of Personality. The result is that psychology does not materially assist us in the study of Personality. Personality is, in fact, largely an unexplored subject and requires a discipline to itself as a real factor in the universe. "Characterology" has been suggested as a name for the new discipline, but there are objections to it, and Personology is suggested as a better name. The "Person" is a concept of the Roman law, not of Greek philosophy, and the hybrid is therefore justified.

Personology should begin by studying the biographies of human personalities as living wholes and unities in the successive phases of their development; in other words, synthetically, rather than analytically in the manner of psychology. Through such scientific studies of Personality we shall obtain the materials for formulating the laws of personal evolution and thus lay the foundations for a real science of Biography. We shall thus also obtain the basis of a sound theory of Personality and a proper science of Personology, which, as the synthetic science of human nature, will form the crown of all the sciences and become the basis for a new Ethic and Metaphysic and a truer spiritual outlook.

To begin with, the lives for such holistic study should be carefully selected, and suggestions are made on this point. The discipline of Personology may thus lead to the solution of some of the oldest and hardest questions that have troubled the heart as well as the head of man.

WE may begin this chapter by defining human Personality roughly in the language which we have adopted throughout the preceding discussion. Personality then is a new whole, is the highest and completest of all wholes, is the most recent conspicuous mutation in the evolution of Holism, is a creative synthesis in which the earlier series of material, organic and psychical wholes are incorporated with a fresh accession or emergence of Holism, and thus a new unique whole of a higher order than any of its predecessors arises. In Personality we reach the latest and highest phase of Holism and therefore the culminating problem, which all the preceding discussion has led up to. Personality is the supreme embodiment of Holism both in its individual and its universal tendencies. It is the final synthesis of all the operative factors in the universe into unitary wholes, and both in its unity and its complexity it constitutes the great riddle of the universe. Best known of all subjects of knowledge and experience, nearest to us in all kinships and relationships, our very foundation and constitution, self of our very selves, it is yet the great mystery, the most elusive phantom in the whole range of knowledge. No wonder that some go the length even of denying its existence, and look upon it as a veritable phantasm of the mind. And yet it is the most real of all reals, the latest and fullest expression of the supreme reality, which gives reality to all other reals. Its uniqueness and its incomparability make it very difficult of approach by the usual methods of scientific procedure, and hence it has been avoided by science completely, and by psychology and philosophy to a very large extent. Perhaps our way of approach to it is a more hopeful one. At any rate our procedure will remove the impression that it is a unique and isolated phenomenon, something alone and by itself, a sort of Melchisedek of the universe, without any genetic connections or contacts with the rest of the universe. We approach it as one of a series, as the culminating phase of a graduated movement of which the earlier steps have been already explored. It thus takes its proper place in the great company of the universe, and is no longer to be viewed as a secluded and unapproachable singularity.

Let us first look at the constituents of Personality. Human Personality takes up into itself all that has gone before in the cosmic evolution of Holism. It is not only mental or spiritual but also organic and material. It is a new whole of the prior wholes; the structures of matter, life and mind are inseparably blended in it, and it is more than any or all of them. What that more is we shall consider just now; in the meantime let us look at its constituents and their relations in the Personality.

The most characteristic and certainly the most important constituent of Personality is Mind. Without conscious mind on the human level Personality could not be. And Personality had to await the arrival of Mind, the development of the organ of Mind, before it could start on its unique career. Mind has been the wing on which the human Personality has risen into the empyrean.

The vast and almost overshadowing importance of the mental or spiritual factor must, however, not blind us to the significance of the other factors, which constitute the body or the physical organism of the human person. These physical organic factors are not only essential, but they also contribute most important features to the human Personality. The human Personality as disembodied spirit and devoid of its physical organism would indeed be something utterly different from what it is. Flesh and blood may not be as important as the soul in the total human make-up, but they are essential and they bring something into the pool which is most vital and precious. So much so that the expression "flesh and blood" has become almost synonymous with humanity. What the Greek poet has called "dear flesh" is not only essential to human nature but gives it a quite peculiar and intimate flavour of humanity. Body is not alien and opaque but indeed transparent to spirit. And the body as transfigured by spirit in man is worthy to be the foundation of the most noble and exalted human Personality. The contempt for the body, the conventional degradation of the body do not spring from a true view of human nature. The natural and proper tendency is to look upon the body as clean and wholesome, to rejoice in it as something good and beautiful, to make it twin-sister of the spirit and the embodiment of joyousness and wholesome pleasure. That view of the body finds characteristic expression in Greek literature. It may be a pagan view, but in reality it is the human and true view. It led to respect and reverence for the body, and the culture of the body as a worthy companion of the spirit. This natural and wholesome attitude towards the body was poisoned by the morbid, diseased, religious spirit of a later time which heaped contempt and degradation on the body. Degraded religions from the East, born amid the filth and squalor, the moral and social decay of the Oriental world, invaded the Roman Empire and found a congenial soil in the moral and religious confusion which had set in among Roman society. The decline of the Empire, the ruin which followed the barbaric invasions, the slow but sure decay of Roman civilisation, and the growing spirit of dejection and despair which was inevitable under these calamities made men turn a ready ear to the base superstitions of the East, which outraged the human spirit and degraded the human body into an instrument of evil. Even the pure spirit of Christianity succumbed to some extent to this perversion, and instead of the body being regarded as " the Temple of the Holy Spirit" it came to be looked upon as a fitter tabernacle for the devil. Mediæval civilisation succumbed to and accentuated this horror of the flesh; the monastic ideal with its monkish practices and morbid celibacies bears eloquent testimony to the great fall of the body. The flesh became synonymous with sin. And it was not till the revolution in the human standpoint brought about by the Renaissance that the body came to be in part rehabilitated and once more to be looked upon with something of the old pagan favour. The full rehabilitation is, however, coming only now at the hands of science. Science is building up a new world-attitude, a new attitude towards Nature and all things natural, which is totally at variance with this morbid and unnatural condemnation of the flesh. The scientific attitude is impartial and objective and leads to the view of Nature as clean, wholesome and worthy of the respect which is due to all natural facts. Thus a new spirit of respect and reverence for natural things and processes is arising, and not least for the human body—a spirit which is far deeper and better founded than the old happy-go-lucky, naïve, pagan attitude of the Greek world. It is not a naïve sentimental, but an objective scientific attitude. It means justice and fair-play for the body, as against all theological prejudices and theories based on erroneous views of human nature. And it completely justifies the normal attitude which regards the body as inseparable from the spirit, and as the source of much that is most dear and precious and beautiful and intimate in human existence. Natural relations and affinities, instead of being condemned, receive the sanction of science, and under the powerful patronage and protection of science the simple human can once more hold up its head and without shame or regret give expression to the spirit of glad wholesome enjoyment which naturally wells up from its inner depths. The rehabilitation of the body is not the least of the magnificent services which science has rendered to human welfare. The body is no worse than the spirit, and can be abused just as the spirit can be perverted. Holism is the cure and remedy alike for the abuse and perversion. It is the severance of body and spirit which makes the ignoble use of either possible. Together and in that unity which constitutes the whole they mutually support, enrich and ennoble each other. It is the division in their ranks which leads to their defeat in detail. And hence it is that Holism is not only a theory but should also be a practice. What theory points out as true becomes here an ideal for life. When spirit irradiates body and body gives massive nourishment to spirit, the ideal of the creative whole as the antithesis of evil is realised in Personality.

The language we have just used seems to imply some form of active inter-relation between the elements in Personality. We seem to assume that body and mind must mutually influence each other in the whole which constitutes Personality. The most difficult and important question, therefore, arises how such mutual influence has to be understood. Do body and mind interact with each other, and how can such interaction be conceived in view of the considerations which were set out in Chapter VII? The difficulties of thought, serious as they are, are here no doubt largely increased by the defects of language. As soon as the "whole" of Personality is analysed into its constituent elements, the elements by the defects of thought and language alike come to be treated as different things, which thereafter can be brought back again into relation with each other only by way of an assumed mutual interaction. Thus arise the division and separation of body and mind which form the very source of the evils we are trying to counter and combat. Thus again, on the basis of this division, the separated elements come to be hypostatised into separate entities or substances which are supposed to interact with each other. Our perfectly fair and justifiable attempt at analysing Personality into mind and body has landed us in an inextricable confusion, which has vexed the soul of philosophy for hundreds of years. In Chapter VII I pointed out that it was this substantiation or hypostasis, as individual reals, of elements which have meaning and reality only as elements in a whole, that is the source of the resulting conundrums. The holistic view of Personality, of Personality as an integral whole, and not as a compound of independently real substances, is the only solvent for these difficulties and misunderstandings.

It may, however, be objected that this holistic view with its implied suppression or squashing of body and mind and the disappearance of both in the Personality is not a fair and honest way of meeting these difficulties. Surely, it will be urged, body is a real, a substance on its own merits, and not a mere abstract element in human Personality. It will be pointed out that at an earlier stage I have treated organism as a real whole, as a holistic structure; that the human body is nothing but an organism and that it is not fair in its human connection to condemn organism as having no reality of its own apart from the Person to whom it belongs. I have stated the objection because it opens the way to the explanation I wish to suggest. A living independent organism is in a different position from the human body. The human body is organic, but cannot be considered an independent organism, living in a sort of symbiosis with another substance called mind. Let anybody try to form an idea of a human body divorced from all mental attributes and activities and supporting an independent existence of its own, merely as an organism. It would not be the human body, whose every organ and activity has a mind-ward aspect and implies mental functioning. Subtract mind, and the residue of body must shrink and shrivel into an unimaginable scrap-heap of organic activities. Similarly it is impossible to conceive mind as abstracted from the body. The disembodied soul is just as impossible a concept as the disminded body. Thus it is that the Christian doctrine of the Resurrection has provided the risen soul with a " spiritual body "[1] the lineaments of which can only, however, be discerned by the eye of faith.

Assuming then that body and mind are not independent reals and have meaning and reality only as elements in the one real substantive whole of Personality, the question arises how we have to conceive or understand their mutual relations as elements in that whole. How does mind influence body and *vice versa* in human Personality? All language implies such influence; all experience implies and assumes that mind has an influence on body, and body on

[1] 1 Cor. xv. 44.

mind. In Chapter VII I have tried to show that such influence does not imply a violation of the laws of physical energy on the one hand, and does not necessitate the intervention of a *deus ex machina* like Entelechy on the other. The situation is entirely holistic, and Holism here as elsewhere gives the basis of the required explanation. Let us first consider the alternative views which have been held on the subject.

(*a*) The popular and, I believe, still the common view among philosophers is that of " interaction "; that is to say, that body and mind mutually and directly act on each other. This view, if rightly understood, does not, as I have said, come into conflict with natural laws. But it is open to another very formidable objection. How can direct action of the physical or material on the mental or spiritual and *vice versa* be conceived? The two supposed interacting factors are not of the same order at all; in what way the one can "act" on the other seems not only unintelligible but absolutely inconceivable.

(*b*) In view of this very grave objection, as well as other objections, many thinkers have simply adopted the view that the physical and the mental are two parallel series, which do not act on each other but run parallel to each other, without any attempt to explain the ground of this psycho-physical parallelism. The objection to this is that one series does seem to influence the other, and not merely to run parallel to it. Our consciousness in voluntary action, for instance, does seem to reveal most clearly that our mental state can influence external actions in a particular direction.

(*c*) Others, again, while admitting the difficulty of conceiving how the physical and the mental can act on each other, have introduced a mediating concept or agency to help the difficulty out. Thus, just as the difficulty of conceiving action at a distance has been mediated by the conception of the ether of space as a medium for such action, so a supernatural medium has been assumed to render possible the interaction of such incommensurables as mind and body. Leibniz has assumed a pre-established Harmony as existing between them and in other respects in the universe; Berkeley and some of the Cartesians have assumed that all interaction takes place in God, the divine medium and cause of all happening in the universe.

(*d*) Finally, there is the view, of which Spinoza's may be taken as the type, that the universe is but one Substance, of which both the physical and the mental series are particular and related modes of activity; the causality which connects them may therefore be supposed to reside in the underlying Substance which unites them both. This view is not entirely unlike the immediately preceding one; for the God of Berkeley and others may be taken to correspond to the divine Substance of Spinoza's universe.

These are the main types of views which have been held on this, perhaps the most difficult subject in all philosophy; and to me all of them seem to contain some elements of truth and value. The holistic conception will not only assist us to regard this subject from a new point of view, but it will do justice to the efforts of those who have laboured at this problem before.

In the first place, then, I may point out that the term "interaction" does not seem well-chosen to describe the relations of two such disparate entities as the physical and the mental. Interaction seems to assume a common platform of action, action on more or less the same level. But we have seen that the structures of matter, life and mind are on quite different levels of organisation and inwardness. The one acts inside the other and through the other. To use a metaphor, the mesh of the one is much finer than that of the other; the lower is transparent to the higher structure, which therefore penetrates it and represents an inner activity which was not there before. Action *through* or *inside*, "peraction" or "intro-action," would therefore be nearer the mark than "interaction" in describing the action of elements in wholes with respect to each other. Mind in "volition" is an inner self-direction of the structure of Body, as I explained in

Chapter VII. Body, again, in giving rise to mental "sensation," is simply performing that mutation or creative leap, which we have found at every other stage of Evolution.

In the second place, and what is even more important, the whole is an active mediating factor in whatever action takes place among its elements. We have referred in previous chapters to the metabolic transformative activity of organic wholes in respect of all stimuli or materials which affect them from the outside. The organic whole itself acts creatively, and subtly changes all alien stimuli or material into its own form and structure. It is the very nature of the whole in each particular case to display this inner creative, recreative and transformative activity. We see it not only in the organism but even more conspicuously in psychic wholes. Thus sensations arise from bodily states or affections. Now in all that happens between the elements in a whole this subtle, creative, holistic factor intervenes. Mind and body as elements in the human Personality influence each other because of their co-presence in this creative whole of Personality. The real actor is the Holism in and of which they are but parts and elements. It is this subtle inner metaboliser or creator which makes all the difference. It is not so much a case of mind and body interacting; rather is it a case of holistic Personality dominating the scene where both of them but humbly serve or subserve. It is the Holism in which they both "live, move and have their being" that is the real explanation of whatever happens or appears to happen between them. It is the Third, which is greater than both of them, that really counts in the action or peraction. To me this, or something like this, is the last word in the relations of mind and body, of the spiritual and the physical. It may sound strange and mystic; but it is the simple fact that the whole, in this case Personality, makes all the difference. Just as Kant at last gave the explanation of mental activity by pointing to the central "synthetic unity of apperception," in other words, to the holistic Subject as the pervading dominating factor, so here *all* action of whatever kind, which happens between mind and body in human Personality, is to be traced to and ultimately accounted for by the holistic Personality itself, and its creative shaping of all that happens to or in it. Any explanation which ignores the Personality itself must necessarily miss the mark.

We now see what is the real explanation of Berkeley and Geulincx's appeal to God, of Leibniz's appeal to Pre-established Harmony, and of Spinoza's appeal to Substance, as the mediator of action between the mind and the body in man, between the spiritual and the physical orders in the universe. The whole in each case is the explanation; the whole as Personality in the human case, the whole as organism in the situation of life and energy. All such action is synthetic and holistic in its very essence, and no explanation which ignores the whole and its creative metabolism in such action can be considered satisfactory.

It may be objected that this " explanation " involves an even greater mystery than that which was to be explained. No doubt we are here moving in a world of mystery, but at any rate the mystery is now rightly placed. We have traced it to its source in the Holism which makes and guides the universe and all its unit structures great and small; and particularly to Personality as a form of Holism. Beyond that final source no explanation can be traced. Personality is a mystery, but at any rate we can attempt to locate it in the order and evolution of the universe. In the relations of mind and body Personality is no mere indifferent spectator or passive *tertium quid,* and the explanation of those relations must in the last resort be sought in the creative activity of the Personality itself. When we analyse material structure into its elements, we can practically afford to ignore everything else besides those elements themselves, because the traces of Holism in such a structure are so faint as to be almost imperceptible. When, however, we go on to analyse an organic whole into its elements, we notice at once that there must be something more besides those elements, something commonly called life which holds all those elements together in a living unity. This "something more" we have identified as Holism, and we have explained it as not something additional quantitatively, but as a more refined and intimate structural relation of the elements themselves. When, proceeding yet higher or deeper, we reach psychic wholes, we become even more keenly aware of the presence and unmistakable function, the free creative activity of this holistic something. And when, finally, we reach the level of personal wholes which include all these earlier less complex holistic types, we find all explanations of action, relation and interaction among the elements futile and hopeless, which ignore this deeper relation, this holistic setting, this active creative Holism which unites all the elements into unique wholes. I believe that previous attempts to state the relations of body and mind in the human Personality have largely failed because the holistic character of the Personality itself as the dominating factor in the situation has been tacitly ignored. It is, however, unnecessary to labour the point further in this connection, as the whole trend of our argument in this work goes to emphasise the importance of the holistic factor in all reality, and *a fortiori* in the highest reality of which we are directly conscious, viz. in our Personality. Any explanation which leaves the Personality out of account in these matters is simply like the play of *Hamlet* without the Prince of Denmark.

Let us now approach the same point from another angle. What is the relation of this Personality to our inheritance from our ancestors? The general principles of organic descent were discussed in Chapter VIII, and an attempt was there made to show the intimate activity of the holistic factor in all organic Evolution. In the last chapter, again, it was shown how psychic Evolution differed from organic Evolution in general; and it was pointed out that the difference was principally this, that while in organic Evolution more or less definite specific modes of reaction to stimuli were inherited, in psychic Evolution, on the other hand, a general plasticity of reaction was inherited, an indefinite range of acquiring experience, a vast capacity of learning in the individual life how to react to any particular stimulus which might happen to come along. The animal, therefore, appears with its very limited range of faculties ready made, so to say, and in its individual life learns very little beyond this definite endowment for specific activities. The human person, on the contrary, has in addition to its organic animal inheritance a psychic inheritance which endows it with the capacity for educability, with a capacity for acquiring new experience and learning new ways of acting and reacting, which raises it infinitely above the merely animal phase. The free, creative, holistic activity of mind appears conspicuously in its hereditary transmission, so that our human inheritance does not fetter us, but by its very nature confers plasticity, freedom and creativeness upon us. What we inherit is not a ready-made affair but a wide possibility and potency of moulding ourselves in our lives. In other words, what above all is inherited is freedom, and the capacity of free and self-determined action and development in our individual lives. In our psychic nature we are thus raised above the bondage of organic inheritance.

Now what does this imply? Surely this, that there is something more in us over and above this inherited endowment. The freedom must belong to an agent; the plasticity implies a creative moulder. I inherit various capacities, but my own Personality itself is not inherited, but is uniquely and originally mine. I inherit a definite animal body, slightly different from those of my parents and ancestors; I likewise inherit a mental structure, somewhat resembling theirs, but much less so than my body resembles theirs. But over and above this organic and psychic inheritance there is an individuality, an individual personality, which makes of this double inheritance a uniquely different blend and composition. The flavour of each human person is uniquely and absolutely individual. However similar the inheritance may be, yet I am a new person, a new self-consciousness, a personal centre absolutely distinguished from those who gave birth to me and transmitted their qualities to me. The unique whole, called Personality, is not inherited, however much its constituent qualities and elements have been inherited. And the very character of the inheritance implies a new conscious centre to which they belong, a centre which will organise them freely and creatively into a new unity. Freedom and plasticity belong not to the experience but to the experiencer. It is the personal self-conscious centre that is free and plastic and creative, just as the artist is free and creative, and not the pigments with which he works.

This line of reasoning leads to somewhat curious results. It will be asked whether there is then a fresh creation of Personality at each generation; whether human Personality is original and underived in the sense that it is newly created with each human being. On the holistic theory here put forward there seems no denying this " creationism," as it has been called. And it is best to recognise at once that with human Personality we enter a domain of creationism, a domain where, far more than elsewhere in nature, creation is at work. We have called Evolution creative; we have seen how creative newness enters at every stage of the evolutionary advance. As the process advances this creativeness increases and intensifies, until in the origin of the highest known structures, that is to say in human Personalities, the creativeness rises to a maximum in relation to the inherited materials used in the new structures. Our very conception of Personality is that it is a unique creative novelty in every human being.

From this it must, however, not be inferred that the phenomenon of Personality should be a most uncertain and wildly fluctuating one; that having no roots in the past and being a creative novelty on each separate occasion, it might be expected to rise and fall with quite incalculable uncertainty. Like all the other holistic structures of Evolution, Personality shows a fair average constancy and probably a tendency to rise slowly throughout the generations. And this is only natural considering the general constancy of the inherited materials which go towards its composition, as well as the constancy of the general environment in any nation or people. But it does show much more individual fluctuation than any other structures in the whole range of Evolution. There is evidently no hereditary character in Personality as such; great Personalities arise from generations of the commonplace; and, again, the great Personality may be followed by generations of the undistinguished. There is utter uncertainty in detail, which goes to prove that with Personality we are in the region of contingency and unpredictability, and that it is not possible to formulate for Personality a law of sequence in the generations. In fact Personality may be compared to biological sports. We know that a species which has its origin in some great sport or mutation often shows a markedly fluctuating character and continues to sport and to show great variability among its individuals in many directions. This is true in a superlative degree of Personality, whose sportive freedom is perhaps its most marked general feature. Still even so there is on an average a fair amount of regularity and constancy, with probably a slow tendency to rise in the scale in the passage of the generations. Even this great spiritual sport (as we may call it) may find its law in the end. But at present it is still utterly individual and incalculable. It is, however, not a mere passing accident or freak of Evolution. It is in line with the whole trend of Evolution; it is a crowning phase of all that has gone before, and if to-day it is still vastly variable and fluctuating, that is so because of its youth, because it has had no time yet to develop firm and constant characters; because it is a whole in the making rather than a whole completely achieved. But its immaturity does not detract from either its merits or its claims. It is a youthful God destined to complete mastery over the old regular Routine, and to achieve Freedom, Creativeness and Value on a scale undreamt of by us of to-day.

From the foregoing it will be seen that the theory of

Personality as a real factor is necessary not only to explain the synthetic relations of the constituent elements in the complex human being, but also to explain the peculiar character of heredity on the human level. Without Personality the Body-and-Mind relation in man appears inexplicable; without Personality, again, man's independence of his hereditary bonds and fixtures would be equally inexplicable. In both respects a far-reaching holistic factor in the nature of Personality is at work, which cannot be ignored without making the entire human situation an insoluble puzzle.

There is a third point of view which lays a strong emphasis on the individual holistic factor which underlies Personality. That is the point of view of Psychology and Epistomology generally. Both these great disciplines erect the human Subject into a new centre of orientation for all experience of reality. From the purely biological point of view Personality is merely the highest term of a rising series; but it is not a new and unique point of departure in the universe. Psychology and Epistomology, however, regard Personality from this radical point of view. To them it is the Subject of experience to which all the rest is the Object of experience. The Personality in experience as the subject marches right to the centre of the world-picture; it becomes the key and the measure of all things; to it all things become relative in experience. In the new universe of experience, in the world of Spirit, the conscious self or the Personality becomes the new point of universal reference; the co-ordinates of reference are *its* co-ordinates, as we saw in the last chapter; and without this personal orientation all experience becomes inexplicable and all reality unintelligible. We may indeed say that as soon as Personality appears on the scene, the whole universe becomes reorganised, transformed and almost recreated round it as the new centre; the universe is no longer the same as it was before Personality but undergoes a radical change in the subtle process of human experience. Just as Personality is essentially a new creation, so the world which is its Object in experience is likewise a new creation out of the old materials. The appearance of Personality, therefore, marks a new departure. It is not merely an addition to the universe but involves an organic transformation of it. On this lofty pedestal psychology and philosophy alike place the personal self or Personality; and surely in this apparent anthropomorphism they are right.

But it is perhaps doubtful whether they have fully appreciated the implications of their action. Neither psychology nor philosophy has made much of the Personality except to look upon it as a peg on which to hang the universe. The Personality as a point of reference, the Personality as a great Signpost in the universe appears to them all-important. But in itself, in what it is, in what its uniqueness consists, they have not taken any very profound interest. Now in this they seem to me to have missed the real point, and in consequence they have failed to appreciate the real, as distinct from the merely formal importance of the factor of Personality in the universe.

The treatment that psychology has given to Personality is another instance of this failure to appreciate its real and unique significance. Psychology as a scientific discipline deals with the human mind, not in its individual uniqueness, but in its general character as distinguishing all human beings. The individual within the purview of psychology is the generalised individual, the average individual, not the real individual, but the individual which is the creature of an intellectual abstraction. In its treatment psychology is, of course, only following the general procedure of Science, which is not concerned with the individual as such, but with the common characters of individuals, with the specific type more than the actual individual. Thus if Science deals with a plant or an animal it may be any individual of the particular species under consideration. The individual differences are generally considered negligible, and one individual is for purposes of scientific treatment as a rule the same as any other individual. Science is a generalising scheme and must necessarily ignore individual differences.

Now with human personalities, the individual differences, so far from being negligible, are all-important. Each human individual is a unique personality; not only is personality in general a unique phenomenon in the world, but each human personality is unique in itself, and the attempt at "averaging" and generalising and reaching the common type on the approved scientific lines eliminates what is the very essence of Personality, namely, its unique individual character in each case. The scientific procedure of psychology, inevitable as it is for psychology as a scientific discipline, is not very suitable in respect of a subject so specially individual as Personality. But that is not all. Psychology does not even purport to deal specially with Personality. Its subject is more especially mind, mental activity in its wider sense, the genesis and development of the mental functions in the average human individual. But, as we have seen, mind is merely one particular aspect of Personality. The contribution towards Personality which comes from the organic side is in important respects ignored by psychology. But this contribution of the body is most important; we know from practical experience in our personal lives how important bodily functions and our general physiological state are in the total make-up of the Personality. Our nervous system, our digestive system, above all our reproductive system have the most far-reaching reactions on our Personality as a whole. A little more iodine in the thyroid gland, for instance, may make the greatest difference, not only for the general co-ordination of physiological functions and bodily development as a whole, but for the mind itself, and may even mean all the difference between normal and deficient mentality, between normal and stunted Personality. All this physiological side of Personality, important as it is in its effects, simply falls outside the scope of psychology and is assigned to other branches of science.

The result of this limitation of the province of psychology is that even the mental side of Personality fails to be properly explored and understood. The subconscious Mind is still largely an unexplored territory, and but for the recent pioneering work of the psycho-analysts would have been almost entirely unknown. And yet it will be generally admitted that this province of the subconscious is most important, not only for mental science itself, but more especially for the knowledge of the Personality in any particular case. For most minds, perhaps for all minds, the conscious area is small compared with the subconscious area; and beyond the subconscious area is the probably still larger organic or physiological area of the nervous, digestive, endocrine and reproductive systems, which all concern the Personality most vitally and closely. It is evident that the present demarcation of areas between the various sciences makes it difficult if not impossible to deal adequately with so large and embracive a subject as Personality. Personality is deserving of having a discipline to itself which will not leave it merely in the position of having to be dealt with in a haphazard and incidental way by a number of other distinct disciplines. Hitherto, so far from having a field of its own and being a study by itself, it has been a sort of nondescript annex of psychology. But, as we have seen, the province of psychology is much too narrow and limited for the purpose of Personality; and both its method and procedure as a scientific discipline fail to do justice to the uniquely individual character of the Personality.

It has a third and no less serious drawback as a basis for a discipline of the Personality. The procedure of psychology is largely analytical; it involves an analysis of mental functions and activities, and a detailed study of their several lines of development; and in exceptional cases a perfunctory effort is finally made to view mind or character as a whole. But mostly the last part is either avoided altogether or attempted in such a half-hearted manner as to be of comparatively slight value. Take, for instance, Professor James Ward's *Psychological Principles,* which is not only a standard work but embodies and expands what has become the great classic in psychology in the English language. It consists of eighteen chapters, the first four of which are devoted to a general analysis and description of mental functions, while the main body of the work, consisting of twelve chapters, contains a detailed discussion of the various forms of mental activity, such as sensation, perception, imagination, memory, feeling, emotion, and action, intellection, forms of synthesis in the judgment, intuition, and the categories, belief, and the elements of conduct. The last two chapters only are devoted to the concrete individual and characterology, and form merely a distant approach to the subject of Personality. No one will deny that from a purely psychological point of view the method and procedure of Professor Ward are both proper and unexceptionable. But the necessarily analytical character of psychology largely disqualifies it from being a real foundation for a doctrine of Personality. Psychology has, in fact, a different scope and aim from that which would be natural and proper for a subject like Personality. It is but one of several preparatory studies leading up to the subject of Personality without actually grappling with it.

The result has been that from a psychological or any other practical point of view very little attention has been devoted to the study of Personality. Personality has been the concern of no particular branch of study, and it still awaits a proper treatment of its own as a distinct discipline among other scientific and philosophical disciplines. Its province falls within the large debatable territory between science and philosophy, between theory and practice, which has been very little explored and is still *terra incognita* to all intents and purposes. Its difficulties are immense; from that wide and wild No Man's Land between science and philosophy it rises like some forbidding mountain peak into the heavens; and no daring mountaineer has yet ventured to approach it, let alone to scale its dizzy heights. But beyond a doubt it is going to occupy a foremost place in the attention of inquirers in future. And the time may come when the science of Personality may be the very keystone of the arch, and serve to complete the full growing circle of organised human knowledge. That time is not yet; but I may venture to hope that the assignment of a proper place to Personality in the structural Evolution of the universe, such as has been attempted in this study, will help to direct attention to what is undoubtedly one of the greatest and most important outstanding problems of knowledge.

Professor Ward has suggested that that branch of psychology which deals with concrete individuals, with individuals as persons endowed with character, should be called "Characterology."[1] I am not clear that Characterology in this sense would be the same as the Science of Personality which I am discussing. The term "character" seems to me narrower than Personality, and to refer to the external indicia rather than the inward reality which the term Personality here points to. And in any case Characterology does not seem suitable as a name for the Science of Personality. For these and other reasons, and if a name is really necessary, I would suggest Personology as the name for the Science of Personality, which will not be a mere subdivision of psychology but an independent science or discipline of its own, with its roots not only in psychology but also in all the sciences which deal with the human mind and the human body. As I have just pointed out, it is a border subject between the provinces of Science and Philosophy and will show the influence of both these great subdivisions of knowledge.

Prima facie Personology seems a more suitable name for the science or doctrine of Personality than the cacophonous mouthful "Characterology." But it may be objected that Personology is a Græco-Latin hybrid and unacceptable as such. It may, however, be pointed out that there is a peculiar reason for a term which is not purely Greek but calls in the resources of the Latin language also. For it is a curious fact that Greek philosophy, in spite of its brilliant achievements and its inspired mintage of most of the current coin of philosophy, never rose to a clear grasp of the idea of Personality. Thus it is that there is no term in Greek to express the notion of Personality. *Persona* is a Latin term

[1] *Psychological Principles*, p. 431.

and a Roman idea evolved, like so many other juristic ideas, by the legal genius of the Romans, which was in its way as remarkable as the philosophical genius of the Greeks. *Persona* in the Roman law denoted the legal status of the individual who was by law clothed with rights and duties in his own right; the individual as the carrier of rights and duties in his own right was a *persona;* from a mere individual nonentity he became in law a *persona* and acquired a legal personality. Thus in the classical Roman law a slave, being without legal rights, was a human without *persona*. The developed Roman law came to extend the concept of personality beyond natural individuals to non-corporeal companies and societies which had by law a legal entity and could have rights and duties. Personality thus was a matter of legal status, and denoted the legal dignity and importance of the individual or the group. The clear juristic concept of *persona* was a very good basis for the superstructure of the psychic ethical Personality which has been built upon it.

To the evolution of the modern idea of Personality, Christianity made the most notable contribution in investing the human being as such with a character of sacredness, of spiritual dignity and importance, which implied a far-reaching revolution in ethical ideas. The Roman legal concept thus became blended with the moral sacredness and inalienable rights of human beings as children of God; and philosophy raised the enriched term to the dignity and status of a high philosophical conception. It has been my endeavour to go a step farther and to trace the concept of Personality to its real relationships in the order of the universe, to show it not merely as a juristic or religious or philosophical concept, but as a real factor which forms the culminating phase in the synthetic creative Evolution of the universe. The Roman traced *persona* to the authority of the law. The Christian traced Personality to the fatherhood of God which conferred it on all human beings as a sacred birthright. The Philosopher has translated this religious idea into the universal language of the ethical Reason. Here Personality becomes the last term in the holistic series, a reality in line with the other realities which mark the creative forward march of Holism.

Personality has thus been explained above as personal Holism, as the whole in its human fullness of development. Human personal development thus means the creative synthetic whole in control of all special functions and activities, of all organs and their functions. The activities of the body and the mind do not embrace the whole of personal development. There is more in the central synthetic Personality than an analysis of psychological and physiological functions can explain. Just as in the specialised organs of sense, the underlying basis of sensitivity, the original *sensus communis*, develops *pari passu* with the special senses and co-ordinates and supplements their activities in the sensuous wholes of intuition, so also the central holistic Personality develops *pari passu* with all the specialised mental and bodily functions, and produces out of their deliverances those syntheses and unities which are distinctive of personal experience. All experience, all intuitions, judgments, actions, beliefs and other mental acts are holistic products of Personality. There is no internal chemistry which binds these products together into unities other than Personality itself. In Personality, even more than in the earlier structures of Evolution, the whole is in charge, and all development and activity can be properly understood only when viewed as being of a holistic character, instead of being the separate activities of special organs, or the separate products of special mental functions. Synthesis and unity are of the whole, and not of the parts. Holism is in all personal activity, and is the only basis on which such activity can be properly understood.

What should be the procedure of the new discipline of Personology? It should, of course, take cognisance of the special analytical contributions of psychology and physiology, and of all the other human sciences, individual and social, theoretical and practical. But it should do more. Following the course above indicated, that the Personality is uniquely individual and that this special individual character should not be ignored, it should study the biography of noted personalities as expressions of the developing Personality in each case. Such a study of personal biographies will not only have the advantage of bringing out the individual differences among personalities, instead of blurring all differences in a generalised composite picture of Personality. It will have the further and quite priceless advantage of studying personalities synthetically as living unities and wholes rather than in the analytical manner of psychology and the other human sciences. In biography we have to follow the development of a person as a whole, as a living biological psychical entity, and we are therefore in a position to correct the one-sided abstract generalised results of the analytical procedure of these sciences. The study of biographies as examples of personal Holism, as examples of the development of Personality, will lead to very interesting and important results.

In the first place, we shall thus get the materials for formulating the laws of personal evolution. In the second place, these laws will form the foundation for a new science of Biography which will take the place of the empirical unsatisfactory patchwork affair which biography now mostly is. In the third place, the gradual accumulation of biographical facts and data bearing on personal evolution not only will lead to the formulation of the laws of this evolution, but will give the basis for a sound theory of Personality and a proper science of Personology. Personology as the science of Personality, as the synthetic science of Human Nature, will form the crown of all the sciences and in turn become the basis of a new Ethic, a new Metaphysic, and of a truer spiritual outlook than we can possibly have in the ignorance and confusions of our present state of knowledge. To my mind the basis for all these great developments can only be laid in a new biographical aim and method, which will give us the facts which are vitally necessary for any sound scientific constructions.

The lives for this scientific study as examples of personal

holistic evolution will have to be carefully selected, if effort is not to be largely wasted. There are many types of personality which would not be specially suitable for studying personal evolution. There is, for instance, the type which does not seem to possess an inner evolution. Many distinguished persons appear to be full grown in early manhood and thereafter to undergo no further growth. Their development reaches maturity early in life, and thereafter appears to be arrested. I may mention Carlyle as an instance; his first great work, *Sartor Resartus,* was a complete and final exposition of his inner self, and no further development of his inner life seems to have taken place thereafter. This phenomenon of early maturity and arrest of further development is by no means unusual. We meet it in the case of many persons in our circle of acquaintances who somehow don't seem to grow, but to stand still after arriving at a certain comparatively early age. We meet it again in the tragic case of those authors who write a famous book early in life and thereafter can do no more than repeat themselves with less and less freshness and ever-waning originality. All these instances simply point to arrested development, to the absence of a capacity for inner growth. As Personality is best studied in its genetic development, as its plastic inwardness is best seen in the successive phases it assumes in a continuously growing, expanding human being, it follows that the exceptional stationary or early maturing personalities afford less favourable material for the study of human Personality as a whole.

There is another class of persons unsuitable for our purpose, consisting of those who do not seem to have much of an inner self at all, whose activities and interests are all of an external character, who live not the inner life of the spirit but the external life of affairs. We often notice this feature in the lives and characters of public men, men of affairs, administrators, business men and others, whose whole mind seems to be absorbed by the practical interest of their work. In them the capacity for the inner life seems to have shrivelled and atrophied under the pressure of external duties and activities. They may be able, competent, conscientious men, they may even be brilliant men of affairs, with great gifts of leadership. They may be striking and impressive personalities and seem to be specially endowed with that indefinable attribute of Personality for which we are searching. And yet they are lacking in that inwardness, that inner spiritual life which is the most favourable medium for the study of Personality. Their lives are generally an affair of externals, of incidents and achievements, sometimes of pomp and glory, but largely devoid of real deep personal interest. Their biographies are usually dull and uninspiring, and the record and recital of activities, successes and failures soon palls on the reader. The fact is that the real indefinable quality of true Personality is inward and is not reflected in the life of unrelieved externality which such people live. They usually carry on the affairs of the world with great competence; but they are too much of the world. What is worse, they often consciously suppress the life of the spirit; the still small voice is no asset to them in the prosecution of their worldly affairs. And they are far too cautious and reserved to give their inner selves away and to afford the outside world glimpses into the world of real motives influencing and guiding them. For them any self-revelation would be something to be shy of, would be like wearing their hearts on their sleeves. The result is that the inner fires are securely banked, and the flame of the spirit can only fitfully smoulder under the ashes. Even if there is a strong personal life in such cases there is usually no record of it, it remains entirely private and personal, and often unknown even to the inner family circle, let alone the scientific student who is dependent on written records, constituting a continuous revelation of the spirit, for the reliability of his conclusions. They may be and often are people of outstanding personality, but the absence of the inner life and of records of personal development make them unsuitable material for the study of the problems of Personality in its more significant aspects.

These remarks will serve to explain what sort of lives could be studied to best advantage in the exploration of the secret of Personality. We should, at any rate to begin with, select the biographies of people who had real inner histories, lives of the spirit, as well as a fair capacity of continuous development during their lifetime. And among these the most helpful cases would be those where the written record is fairly full in the form of writings and diaries, and where there was no undue restraint in the process of self-revelation and faithful portrayal of the inner life and history. On the whole, the lives of poets, artists, writers, thinkers, religious and social innovators will be found the most suitable for purposes of holistic study. They are often people with inner lives and interesting personalities, with an inner history of continuous development; and wherever their experiences have been more or less faithfully recorded, the materials for fruitful study are present. Sometimes the personal record is missing, and in such cases the study of the Personality through the works of the author becomes too much a matter of inference to be really useful, at any rate in the earlier stage of the inquiry into Personality. Such a case, for instance, is that of Shakespeare. His plays reveal behind them a wonderful Personality endowed with the highest genius, and moving forward in a continuous grand crescendo of self-development as an artist from beginning to end. But while the development is there, the Personality itself is too much hidden behind the dramatic mask, and therefore too much a matter of inference in the absence of proper personal records. In other cases, again, the personal record is well and fully known, but the written works are not sufficiently illuminating as a true index of the growing Personality. For a man often reveals himself more profoundly in his master products than in his diaries or correspondence or other incidental communications. Milton's great dictum holds for all time: "A good book is the precious life-blood of a master spirit, embalmed and treasured up on purpose to a life beyond life." There is nothing trivial in Personality, and the greatest, most serious work is usually the most faithful index to the Personality behind. Both are, in fact, required—the work as well as the personal record—for a full understanding of any particular Personality.

From a series of biographical studies, such as I propose, it will, I imagine, become clear that Personalities follow their own laws of inner growth and development, which will, while conforming to a general plan, show very considerable diversity in detail. It will be found that each Personality is a psychic biological organism, an individual personal whole, with its own curve of development, and its own series of phases of growth. A person will thus be found to be very different at different stages of his development, but all the stages and phases will be bound together by and be the outcome of the identical inner Personality. A comparison of such studies of individual Personalities will then give the curve or the law of Personality, and reduce to rational order a phenomenon which is to-day still within the region of mystery.

As the key to all the highest interests of the human race Personality seems to be quite the most important and fruitful problem to which the thinkers of the coming generation could direct their attention. In Personality will probably be found the answer to some of the hardest and oldest questions that have troubled the heart as well as the head of man. The problem of Personality seems as hard as it is important. Not without reason have thinkers throughout the ages shied off from it. But it holds precious secrets for those who will seriously devote themselves to the new science or discipline of Personology.

CHAPTER XI

SOME FUNCTIONS AND IDEALS OF PERSONALITY

Summary.—The central conception of Personality is that of a whole; it is the most holistic entity in the universe, hence no other category will do justice to it, and certainly not mechanism. Psychology is too much of an abstract science to give an adequate view of Personality, though even psychology is dependent on the theory of a central synthetic activity for the correct construction and interpretation of mental experience, and ignores that theory at its peril. The suggestion of a new science or discipline of Personology has therefore been made which will study Personality more synthetically and concretely than is possible for psychology.

As an active living whole, Personality is fundamentally an organ of self-realisation; the object of a whole is more wholeness, in other words, more of its creative self, more self-realisation. This means that the will or active voluntary nature of Personality is its predominant element, and the intelligence or rational activity is subordinate and instrumental—it has to discover and co-ordinate means to the end of self-realisation. Feeling is likewise subordinate, its function being to give strength and impetus to the will. The Personality is thus a more or less balanced whole or structure of various tendencies and activities maintained in progressive harmony by the holistic unity of the Personality itself. In fact Personality resembles an organised society or state with its central executive and legislative authority wielding sway over its individual members in the interest of the whole. Kant has rightly called man a legislative being. Part of this control in Personality is conscious, most of it is, however, subconscious. This control is still largely imperfect and immature owing to the extreme youth of Personality in the history of Evolution. But it is growing. More holistic control in the Personality means greater strength of mind and character, better co-ordination of all impulses and tendencies; less internal friction and wear and tear in the soul, more peace of mind, and finally that spiritual purity, integrity and wholeness which is the ideal of Personality. The Personality has the same self-healing power which we saw already in the case of the mutilated organism; and in case of moral or other aberration it usually has the power to right and recover itself and often creatively to gather strength from its own weakness or errors.

Personality is not only a self-restorer; it is a supreme spiritual metaboliser; it absorbs for its growth a vast variety of experience which it creatively transmutes and assimilates for its own spiritual nourishment. As metabolism and assimilation are fundamental functions of all organic wholes, the Personality takes in and assimilates all the social and other influences which surround it, and makes them all contribute towards its holistic self-realisation. Personalities vary greatly in their capacity for holistic assimilation, some easily suffering from spiritual indigestion, while great minds and characters can absorb a vast experience which only serves to fructify and enrich them without any detriment to their spiritual wholeness and integrity. Where a Personality takes in alien experience which it cannot assimilate into its own spiritual substance, such experience becomes an impurity to it; "purity" in reference to Personality meaning the absence of all elements alien, heterogeneous and disharmonious to the Personality.

The holistic categories sketched in Chapter VI are specially characteristic of Personality as a whole *par excellence:* these are Creativeness, Freedom and Wholeness or Purity. Its creativeness refers to the ideal Values, rational, ethical, artistic and religious, which it creates for its own spiritual environment and inner guidance and illumination. As these, however, fall outside the scope of this work, the category of Creativeness as applying to Personality will not be further considered here. But something must be said about Freedom and Wholeness or Purity.

The essence of Personality is creative freedom in respect of its own conditions of experience and development; as an initiator, metaboliser and assimilator it has practical self-determination. Again, as a selector and co-ordinator of the elements in the situations that confront it, it also has practical freedom. Its very nature as a whole confers freedom upon it. This freedom is not a negation of the physical order of causality but arises inside that order; holistic freedom is a continuous organic or psychic miracle which happens *between* cause and effect, so to say, as we saw in Chapter VI. Freedom is thus a fact in the universe, and is not a mere capricious power peculiar to the will; it pertains to Personality as a whole. Freedom means holistic self-determination, and as such it becomes one of the great ideals of Personality, whose self-realisation is dependent on its inner holistic freedom.

As regards Wholeness or Purity, it is essentially identical with Freedom. Purity means the elimination of disharmonious elements from the Personality. It means the harmonious co-ordination of the higher and lower elements in human nature, the sublimation of the lower into the higher, and thus the enrichment of the higher through the lower. From this it follows that moral discipline is an essential part in the culture of Personality. Personality is a spiritual gymnast, whose object is the freedom and harmony of the inner life through the refinement and sublimation of the cruder features in the Personality and their subordination and co-ordination in the growing whole of Personality. If this object is secured by the Personality, all the rest will be added unto it: peace, joy, blessedness, goodness and all the great prizes of life. Wholeness as free and harmonious self-realisation thus sums up the *summum bonum* of Holism.

IN the preceding chapter we have viewed the Personality as a whole, as a form, and indeed the highest form of Holism; and we have also considered some of the difficulties and problems which arise from this view of Personality as a real whole. In this chapter we shall consider Personality in action, in its operation as a whole, as an active shaping factor in the life of the human individual. Let me, however, first briefly resume what was said about the holistic character of Personality, especially in its psychological aspect.

In Chapter VII I tried to reconcile the conflicting claims of Mechanism and Vitalism in the larger setting of Holism. In considering the behaviour of organisms and organic control generally, it may still be a question whether the Mechanistic or the Vitalistic aspect of Holism is predominant; when, however, we come to the conscious human Personality the question loses all its force and meaning. For there can be no reasonable doubt that the mechanistic conception is not competent to explain or even describe the facts of human Personality. Psychology itself is unintelligible except on the assumption that in Mind we have a central synthetic power which marshals and controls, and largely determines all the facts and functions of mental life, such as sensations, perceptions, conceptions, conations and emotions. Our developed consciousness directly reveals an identical and persistent Self which refers all its experiences to itself; and, as we have seen, but for such a personal centre and unity of reference, mental life and experience would be impossible and unintelligible. This personal Self underlies, upholds, directs and controls all our experience as individuals. In this Self we behold, not only what is deepest and most central in ourselves as human beings, but also that power of Holism which operates blindly as life and organic control in organisms; nay, more, in it we behold the culmination of that fundamental holistic motive power of the universe, the beginnings of which lie far back, impersonal and embedded in the inorganic order of Nature, but which gradually disentangles and frees itself, until in the Self of the human Personality it attains its highest measure of freedom. The synthetic organising power of Holism, starting from the darkest and feeblest beginnings and blindly battling with all sorts of refractory situations in the course of cosmic Evolution, gradually evolves and wins through until at last it emerges in the Self with luminous and radiant self-consciousness. Through the Self, which possesses the power of conscious reflection and retrospection, Holism can look back to its own early beginnings and review its own progress throughout the course of organic and inorganic Evolution. As Nature finally learns to read herself with the human eyes which are her own, so through the human Self which is the highest and best it has yet come to, Holism may gaze back to its beginnings and scrutinise what would otherwise be dark and unintelligible for ever. And thus it is that the worm of Personality comes to turn to the light of the Whole, and presumes to view and discuss the Whole, of which it forms itself but a part.

As was pointed out in the last chapter, the procedure of psychology is largely and necessarily analytical and cannot therefore do justice to Personality in its unique wholeness. For this a new discipline is required, which we have called Personology, and whose task it would be to study Personality as a whole and to trace the laws and phases of its development in the individual life. Such a study would be of the greatest interest from every point of view, as it would envisage Personality in its unique wholeness and unity, rather than, in the way of psychology, as a series of separate abstracted activities. Personology would study the Personality not as an abstraction or bundle of psychological abstractions, but rather as a vital organism, as the organic psychic whole which *par excellence* it is; and such a study should lead to the formulation of the laws of the growth of this unique whole, which would be not only of profound theoretical importance, but also of the greatest practical value. One cannot read the lives of the great Personalities without feeling that a vast field for first-class scientific and philosophic research remains still unexplored, and that discoveries of the highest importance await the student of Personology. Here I shall confine myself to a few indications of the general activity of Personality as a whole.

As a whole, as the individualising power and activity of Holism, the Personality is fundamentally an organ of self-realisation. As in the case of the growing or mutilated organism the whole manifests itself by bearing through all obstructions and overcoming all obstacles in its efforts to realise and complete itself or its type in each individual case, so too the Personality has, as its central end, the straightening out of all difficulties and the elimination of all elements which militate against the attainment of its own immanent ideal. In essence the task is the same in both cases. But there is this material difference in objective, that whereas in the case of organism the end towards which the whole is moving is the completion of the material structure and its functions, in the case of Personality, on the other hand, the end and object of the inner whole is the realisation of an invisible spiritual structure or character. The organism is still mainly material, while the Personality is essentially an inward ideal; but in both cases the shaping power of the inner whole strives to realise its end, to eliminate what is alien and adventitious, to conserve and develop what is pure and relevant to its ideal, and so to reach perfection, of visible outward structure and function in the one case, of inward spiritual grace and unity in the other.

From this it will be seen that apart from our bodies the basis of that complex whole which we call the Personality is our voluntary activity or the will; it is the active, self-maintaining, self-realising power of the Personality in us which underlies and directs and to a large extent conditions all other activities. The Intelligence has been evolved largely though not entirely as an instrument of the will; in its endeavour to realise its conscious or unconscious ends the Personality *qua* will has developed the intellectual or thinking power as a subsidiary activity which prescribes the means by which that realisation has to be effected. The power of Holism in us moves at first unconsciously and blindly, as in other organisms, and later on consciously and purposively to certain ends which increase in complexity and difficulty as the capacity for abstract thinking and rational co-ordination progresses. This fundamental movement is the will, whose activity is dependent not only on the primary forms of feeling, which make the movement slow or rapid according to the strength and volume of the feelings, but also on the growth of intelligence which adjusts means to ends. The active movement to satisfy the appetite or craving of hunger, for instance, depends largely on the strength of the promptings of hunger; and the intelligence of the hungry animal is developed and sharpened in order to devise ways and means by which the pangs of that craving may be alleviated and removed. And similarly the complex impulse which makes a great thinker, artist or statesman endeavour through long years to execute some great and far-reaching plan, while fundamentally a movement of his active voluntary nature, depends for its strength on his emotions, and for the correctness of execution on the power of thought and judgment and insight which have been matured in the personal life. The conception of Personality, as an active movement of the whole in each individual, seems, therefore, necessarily to lead to the primacy of the will or active nature of the mind, and to the instrumental character of the intellectual or thinking power. Personality is thus a balanced whole or structure of various tendencies and capacities which are maintained in mutual and reciprocal harmony by the holistic nature of the Personality itself. As the whole is the essence of Personality, so wholeness in self-realisation and self-expression is its essential aim and object.

The great practical problem before the Personality is thus to effectuate and preserve its wholeness through the harmonising of its several activities, and the prevention among them of any random discord or sedition, whereby one or other might be enabled to assume ascendancy over the rest and so prepare the way for the disintegration and destruction of the whole. In the Personality there is superadded to the unconscious organic control a whole complex machinery of conscious purposive action which is intended more effectively to maintain and increase this highly organised harmony in the developing individual. The machinery of conscious purposive control becomes highly elaborate and almost artificial.

In fact the nature of the Personality is distinguished by its departure from the processes of organic nature and an approximation to the forms of action which are characteristic of society. Just as in a well-organised society or state there is a central legislative and executive authority which is for certain purposes supreme over all individuals composing that society or state, and controls their activities in certain definite directions deemed necessary for the welfare of the state, so the human Personality is distinguished by an even more rigorous inner control and direction of the personal actions to certain defined or definable ends. This is the reason why Kant has called man a legislative being. He is an inward kingdom or sovereignty, whose powers and actions are directed, not by some external agency, but by an inner agency which is none other than the activity of the personal whole itself. Much of this control and direction is conscious will, but far more is unconscious and operates in the subconscious field of the personal life, and it is only on great occasions or crises that light comes suddenly to be thrown on this inner leading in the personal life, and the individual becomes conscious that he has been guided or led along paths which were apparently not of his choosing, but which nevertheless were the outcome of the mysterious inner self-direction which distinguishes the Personality. The ideal personality is he in whom this inner control is sufficiently powerful, whether exercised by conscious will or some unconscious activity, to harmonise all the discordant elements and tendencies of the personal character into one harmonious whole, and to restrain all wayward, random activities which are in conflict with that harmony. This ideal is far from being realised universally in practice. Personality is still a growing factor in the universe, and is merely in its infancy. Its history is marked by the thousands of years, whereas that of organic nature is marked by millions. Personality is as yet but an inchoate activity of the whole, but nevertheless its character is already distinct and well-marked; and its future evolution is the largest ray of hope in human, if not terrestrial, destiny. Its incomplete imperfect character is largely responsible for the interminable disputes and differences among philosophers and theologians about the human soul and human destiny. For so long as the true nature of Personality, which is one form or another and whether consciously or unconsciously, forms the ultimate subject matter of all their dogmas and speculations, is still indefinite and undetermined, it is not to be expected that they will be agreed as to the fundamental postulates, or the proper methods to be followed, or the correct inferences to be drawn from the apparent facts. The scientist has the advantage that he deals with older well-marked manifestations of reality about whose definition and principal characteristics there can be little dispute. But philosophers, whose subject matter is still in process of growth and inward definition, find themselves unable to agree about fundamentals largely because Nature herself is not yet certain about these fundamentals. However, even admittedly inchoate and infantile as Personality is, it is already sufficiently developed and distinct to enable us to consider its fundamental characteristics and their bearings on the interpretation not only of human conduct but of our conception of the universe in general. And its fundamental character is just this wholeness which justifies us in saying that Personality is a special activity or form of the Whole. For consider for a moment what distinguishes the formed and developed personality from the unformed and incomplete personality; the strong character from the weak; the master of his fate from him who is blown about by every wave of impulse or opinion. In the latter case—the case of the weak, or flabby, or irresolute person—you have usually the same elements of character as in that of the strong man. But the difference is that while in the case of the strong man or personality all these elements are unified into one central whole which shapes and directs their separate activities, in the case of the weak man these elements of thought, emotion, will and passion have never been harmonised or fused into one whole; the sovereign legislative and executive authority in the personality has never been properly constituted or exerted, or is so weak as to be regularly disobeyed and defied; the unorganised and uncoordinated factions in the character fight for their own hand and keep up a constant state of inner warfare in the personality, with the result that the stronger passions or impulses carry the day and ruin the character, which depends on a harmonious subordination of all the various elements of character under one supreme ethical authority. The inner discord may even proceed the length of apparent dissociation of the personality and lead to the singular phenomenon of multiple personality in the same individual.

In proportion as a personality really becomes such, it acquires more of the character of wholeness; body and mind, intellect and heart, will and emotions, while not separately repressed but on the contrary fostered and developed, are yet all collectively harmonised and blended into one integral whole; the character becomes more massive, the entire man becomes more of a piece; and the will or conscious rational direction, which is not a separate agency hostile to these individual factors, but the very root and expression of their joint and harmonious action, becomes more silently and smoothly powerful; the wear and tear of internal struggle disappears; the friction and waste which accompany the warfare in the soul are replaced by peace and unity and strength; till at last the Personality stands forth in its ideal purity, integrity and wholeness. And through all this transformation from the disorganised atomic state to the full realisation of unity in the personal character, the Personality as the activity of Holism in the human individual is itself the creative shaping agency which directs the movement; it is the Personality which not only develops all the separate faculties of mind and soul, but concentrates and finally unifies their activities; the various mental elements it organises and fuses into one luminous personal whole, which in time exercises a restraining and overshadowing power over all tendencies and impulses harmful to the whole, and directs the entire current of being, thinking and feeling to the realisation of the highest ethical and spiritual ends.

We have seen in earlier chapters how in case of mutilation of an organism some central control often avails to restore or repair the mutilated organ. In the same way the Personality, as an activity of Holism in the individual, repairs any breach in the personal character and restores the balance disturbed by any impairment of character. The Personality appears as the self-healer, which through all obstacles and impediments endeavours to preserve and realise its own type or ideal, and often even from defeat and disaster itself to wrest the accomplishment of the ethical ideal at which it is consciously or unconsciously aiming. Not seldom, of course, the Personality finds it impossible to overcome the defeats it has sustained and goes under; for it is as yet weak and inchoate as a function of Holism, and in some cases it is weaker than in others. But the level of its power and activity is gradually rising; more and more it is gathering the unorganised centrifugal tendencies of the individual into an effective central control, and often it wins even in the most discouraging circumstances those moral victories which form the great landmarks of personal and human progress. From the depths of moral and spiritual aberration it guides the weak steps of the wanderer to conscious manhood and moral self-control. As the organism heals itself after a mutilation, so the Personality through identically the same functioning of Holism saves and purifies the personal character often even by means of the sins and excesses of which it has been guilty. Thus the Personality realises itself by producing unity and wholeness in the personal character; and when through its own weakness the character is degraded and a course of conduct embarked on which constitutes a denial of that fundamental tendency and aspiration towards wholeness, the force of the Personality in the individual is often strong enough to rescue the individual and sometimes even through a more or less violent crisis to convert him to sanity, self-respect and moral wholeness. The moral and spiritual implications of this fact lie beyond the scope of this work.

The aberrations of the individual from the ethical standard are due not only to the inner weakness of the personal character but also to the influence of the environment. From the consideration of the internal we therefore pass on to discuss the external relations of Personality. And here the first point to note is that in so far as the individual is at the mercy of external circumstances and forces, the situation is largely of a mechanical character. We have seen in earlier chapters how such a mechanical situation is converted into an organic one. The organism does not merely passively receive the force, pressure or influence of the environment; it appears not as a mere passive sufferer, but as an active agent in the drama of existence. And it is considered an organism only to the extent to which it exercises this active function of assimilation or metabolism of the material which it receives from the environment. So far from being a mere channel or conduit pipe for transmitting the inorganic forces and energies of Nature, it disintegrates all the materials supplied to it, and transmutes them into forms which are serviceable for its own organic purposes, and then builds these materials so transmuted into the stately type which it is its immanent end to realise. The power of assimilation is essential to the organism; without this power it would simply be flooded with its surroundings, and instead of conquering the environment and victoriously adjusting itself to its surroundings, it would be overcome and disappear as an organism. Metabolism and assimilation are indeed the fundamental activities of organic wholes.

Now all this is, *mutatis mutandis*, even truer in relation to the Personality. Any element of a foreign, alien or hostile character introduced into the Personality creates internal friction, clogs its working and may even end in completely disorganising and disintegrating it. The Personality, like the organism, is dependent for its continuance on a supply of material, intellectual, social and such-like, from the environment. But this foreign material, unless properly metabolised and assimilated by the Personality, may injure it and even prove fatal to it. Just as organic assimilation is essential to animal growth, so intellectual, moral and social assimilation on the part of the Personality becomes the central fact in its development and self-realisation. The capacity for this assimilation varies greatly in individual cases. A Goethe could absorb and assimilate all the science and art and literature and take part in much of the practical administrative activities of his time and place without finding himself oppressed by a load which must have killed a lesser man; he could, as he has described in the character of Faust, gather up into himself not only all the knowledge of his day, but all the richness and variety of experience which makes his life one of the most interesting records in the history of the world; he could drink of the deepest fountains of passion and arise to the loftiest heights of ideal aspiration—he could do all this and not only preserve his spiritual manhood unimpaired, but actually deepen and broaden and enrich it in every direction. He could assimilate this vast mass of experience, could make it all his own, and make it all contribute to that splendour and magnificence of self-realisation which has made him one of the greatest among men. A lesser Personality would have gone under; either could not have acquired so much knowledge and experience, or could not have assimilated it, and in the end would have become depersonalised, a mere mechanical acquirer and hoarder at the cost of essential unity and integrity. As soon as a person acquires either knowledge or experience or falls under social or other influences in a mechanical manner without assimilating them, he injures his Personality; he overburdens and disorganises himself; he surrenders to the environment that in him which is and should ever remain a pure unconstrained self-activity. There are many forms which this enslavement of the Personality takes. Looking upon the Personality as merely a natural activity and not yet in an ethical or religious light, we find that it is sometimes overloaded and gorged with knowledge which it cannot assimilate and digest, and the person degenerates into a mere gatherer of knowledge, a sort of intellectual hoarder. In other cases, again, it accepts the social influences and conventions without mastering and assimilating them and develops into a purely conventional character in which the spontaneity of the inner life is deadened under a mass of social conventions. In other cases it acquires power which is beyond its capacity to use wisely and well, and it develops a proud, cruel, overbearing or tyrannical character, and that too under circumstances which would have built up a strong and noble Personality in a case where the assimilative, controlling, co-ordinating power was greater. Too often, alas! it simply surrenders itself weakly and self-indulgently to outside influences or temptations, and becomes weak, vicious and contemptible. In all these cases the Personality succumbs to the environment, to external influences which bear on it, but which it cannot resist or master and make its own; in fact, to the introduction of foreign or hostile material into its pure inner self-activity. The ideal Personality is a whole; it is a whole in the sense that it should not have in it anything which is not of a piece with itself, which is alien or external to itself. Any such extraneous or adventitious element in it which does not really harmonise with it prevents it to that extent from being a whole. Now as the Personality is a self-realising holistic activity in us, it follows that its immanent end and ideal is to realise and develop itself as a whole, to establish and secure its wholeness, and to render itself proof against invasion and injury from all extraneous and hostile influences. It cannot do this by cutting itself off from the environment on which it is dependent for the material which it requires for its sustenance and self-realisation. It can do this only by, on the one hand, developing and strengthening its power of assimilating and making an integral part of itself all the materials which are necessary for its requirements, and, on the other, rejecting all unassimilated extraneous materials which come to it without being incorporated into it as a whole. In other words, it aims at *efficiency* and *purity*—the assimilation or making its own of whatever is required for its self-development, and the rejection of all influences or materials which are extraneous to its wholeness, which would be alien and impure to that wholeness.

The term "purity" is here used in the same sense in which the German *"Reinheit"* is often used, to indicate the absence of matters or influences which are alien or non-homogeneous or extraneous to the thing in question. A thing is called pure when it is free from such alien or extraneous or adventitious elements as are considered destructive of its integrity and simple transparency or homogeneity. This seems to be the fundamental meaning of the term "purity." Its moral meaning as freedom from vice, or hygienic application as cleanliness or freedom from dirt are essentially derivative. If an object is itself and nothing but itself, without the adherence of any adventitious matter foreign to it, it will be pure or clean in the fundamental sense. If a person keeps out of his nature any warring or jarring elements or complications, keeps himself free of all moral or spiritual entanglements, and is nothing but himself —whole, simple, integral and sincere—he will also be pure in the vital holistic sense. The food which enters the organism as alien material is destroyed as such in the process of metabolism and is assimilated as blood and other substances and goes to feed the organic system and to form an essential part of it. And similarly the Personality through perception, intuition, conception and emotion, etc., assimilates the influences of its environment and works them up into its own substance—its inner world of thought and will and emotion. And the more thoroughly this mental or personal assimilation is carried out, the richer and more distinctive the Personality is. The wider the range of its acquisitions, the more powerful and thorough the intellectual and emotional assimilation, the more complex and the grander is the Personality.

Here then we reach the central idea and function of the Personality. Like organism, only in a far more complex and developed form, it is a whole, with an interior conscious self-direction of all its component functions; with a power of acquisition from its environment which is not mechanical, but really transforms all the acquired material into transparent unity with its own nature. It is a whole which in its unique synthetic processes continuously performs that greatest of all miracles, the creative transmutation of the lower into the higher in the holistic series.

And, in order to maintain the right perspective, let us not forget that Personality is but a specialised form of Holism, this Personality in all its uniqueness is still but a function of Nature in the wider sense; that in it we see matter itself become somehow aglow and luminous with its own unsuspected immanent fire; that as Personality transforms the material into the spiritual, so regressively a deeper view discloses Personality itself as but a more interior function of that Holism which has been slowly evolving since the beginning of the universe.

In fact Personality in its fundamental activities illustrates all those functions which in Chapter VI we have ascribed to wholes. As a whole it is creative, it is free, and it is unified in the highest sense. In that chapter the groundwork of the holistic categories was laid down, and those categories themselves were derived from the concept and nature of wholes. Personality is the highest type of whole which we have knowledge of, and we should therefore expect to find that the holistic categories of Creativeness, Freedom and Wholeness will apply in a pre-eminent degree to the functions and activities of Personality. I shall conclude this chapter with a brief statement of Freedom and Wholeness or Purity, as illustrated by Personality. The category of Creativeness in its full application to Personality is best illustrated by the appearance of the great artistic, ethical and spiritual Values or Ideals, which are the creations of Holism on the personal plane. These Values and the higher world of the human spirit which they constitute fall beyond the scope of the present work, which is concerned more with the laying of the foundations of the holistic concept than with the erection of the superstructure. The creative Ideals of Holism in their human aspects, although they give better illustrations of Holism than anything we have discussed in this work, will not be dealt with at this preliminary stage. I therefore proceed to discuss Freedom and Purity in their application to Personality.

The creative power by which both organism and Personality metabolise and assimilate extraneous materials raises the issue of Freedom in an obvious and natural way, and we may briefly resume here what has been said before as to the rival claims of Freedom and Mechanical Necessity in their application to organic and personal wholes. In Chapter VII I have explained in what sense and to what extent the categories of Mechanism and Necessity apply to such wholes. That to a certain extent they are mechanisms falling within the physical laws of Necessity is clear; but only to a certain extent. Beyond that Holism appears as a real active factor in each such whole, controlling and directing its physico-chemical energies towards definite ends.

The free activity of Holism in the organism and in the personality, considered merely as an organism, does not affect the mechanical chain of natural causation. In an organism the same combination of physical causes produces the same total of physical effects as in any other system. As we saw in Chapter VII, the law of the conservation of energy holds exactly in the same way as in any other natural system. Holism does not break the causal chain; it does not override the laws of physical causation. The laws of physics and chemistry are the same, whether they are studied in the growth of a crystal or in the development of a plant or animal. To that extent and in that sense Necessity reigns in the plant or animal no less than in the crystal. But that does not exhaust the matter. On the basis of these natural conditions and factors Holism proceeds to bring about results which are impossible in the case of mere mechanisms. Holism does not annihilate its form of space when it proceeds on its road of inward development, but within the limits and limitations of the spatial external form it proceeds to the creation of a new inner world. In the same way Holism accepts its own well-known natural conditions and principles of action when it comes to develop inward organic or personal wholes, but it evokes meanings and values and results from those conditions which would have been impossible on the plane of the merely spatial or mechanical. Holism, while in no sense overriding natural factors which are but an earlier phase of its own activities, develops inside and through those factors the individual wholes of organism and Personality. Similar causes produce similar effects under similar conditions: that is a statement of natural law. But the miracle of Holism is performed in that infinitely small or timeless, spaceless interval which elapses between cause and effect. Hence whereas on the physico-chemical plane cause A is followed by effect B, in the case of an organism the operation of Holism is seen in that cause A is followed not only by effect B, but also by a new non-mechanical element X of a holistic character in the shape of what is ordinarily called life or sensation, organic or mental activity. Organism as a whole is not merely a link in the chain of natural causation, but is itself an absorber, assimilator and transformer of causes on the way to their effects. And this active free power of absorption, assimilation and transformation is evidenced not only in the creative appearance of the new vital or mental element X, but also in the natural sense of freedom which accompanies this activity in personal consciousness. A causal stimulus applied externally to an organism does not merely result in some mechanical movement, but between the stimulus and the resulting movement a whole new world intervenes, which transforms the stimulus into the state of the organism, and makes the resulting movement, not the mere mechanical effect of that cause, but the free action of the organism. The organism absorbs the cause as mere material, and emits the movement as the resulting action of itself as the real cause. This transformation not only is seen to happen in the case of the lower organisms, but is revealed and interpreted in human consciousness as what actually does take place. Consciousness interpolates the self between all causal stimulus and all resulting response, and reveals the self as the free creator or prompter of the response after it has absorbed the stimulus. Accordingly, as we saw in Chapter VI, freedom arises creatively inside the process of natural causation.

The spontaneous self-activity of the organism in the assimilation of material necessary for its nutrition and development shows that it is free as an organic whole; while the assimilation and transformation of that material and the reference of any resulting movements or responses to the organism as their originating and determining cause show that freedom or self-determination from another point of view. There is no such spontaneity nor such power of creative assimilation in any mere mechanical aggregate; in so far as an organism is a whole, it is also a free self-determining agent in the activity which dissolves and assimilates extraneous influences or materials and substitutes freedom for causal necessity.

We thus see that Freedom has its roots deep down in the foundations and constitution of the universe. It is a profound mistake to look for Freedom only in the human will. The correct and fruitful view discloses Freedom, not as an exceptional development in the universe, as an attribute merely of the human will, but as itself in one degree or another the grand rule of the universe, as the free self-determined activity of Holism in its universal process of self-realisation in Evolution, and as the fundamental principle of each individual whole set free in the course of this Evolution. As Holism in its individuating activity evolves and sets free smaller wholes, these wholes are themselves in ever-increasing measures set free from external determination and acquire an ever greater measure of self-determination and freedom in their activities and development. Holism not only means the development of the universe on holistic lines, the realisation of ever more perfect wholes, and the assimilation, transformation and absorption of non-holistic material or relations. It means also the ever-widening reign of Freedom, the realisation of the Ideal of Freedom in the gradual breaking down of all external fetters, and the gradually increasing inward self-determination of the universe through the progressive evolution of ever higher holistic entities in the universe. This free holistic activity is not only the source of the idea of causation in human consciousness; it is ultimately the only source of efficient action or causation in the universe. The free activity of Holism or a whole is the type and source of all efficient causation. The concept of necessity, which arises in connection with that of causation, is not grounded in the reality of things, but is (as Kant showed) a mere mental expedient for joining up or reconnecting parts of the whole which have become dissociated or severed in the course of thought or experience. The synthetic activity of mind, in producing the category of necessity, is simply intended to recover or reconstitute intellectually that whole which has been shattered into fragments in experience and thought; and as mind is itself but part of the larger whole of Personality, this intention can only be carried out imperfectly. In the whole Freedom and Causation, or rather efficient action, are not utterly different; their antagonism arises only in the application of consciousness to the atomic aspects of our empirical experiences. Determinism is in the last resort based on free holistic self-determination. We may sum up by saying that Holism is free, and in so far as Holism has realised itself in the universe, in so far as the universe is of a holistic character and consists of holistic entities, to that extent the universe and these entities are themselves free.

But Personality is the highest type of such holistic entities. We may therefore say that Personality as a whole is free; the more completely it realises the character of a whole, the more perfect also will be its freedom as such. The freedom of the Personality is simply its character of pure self-activity, untrammelled by external influences, its character of spontaneous or conscious self-determination by virtue of which all its actions flow from the pure source of self and are not pressed or forced on it by unassimilated external conditions or causes alien to itself, and which have not been transformed into unity with itself. Sincere self-expression in men and in nations thus becomes the true ideal of human development and culture.

Freedom is thus not a mere abstract formal concept, but a real activity; it is the limits within which Holism moulds and develops the individual Personality. In proportion as the Personality is holistic, it is rich in the characters of self-direction and self-determination; in other words, it is free. Moral Freedom is thus a form of the holistic activity of Personality.

It will be seen that we predicate Freedom, not of the Will, but of the Personality itself. However important and indeed fundamental an aspect of Personality the will is, yet it is merely an aspect and not the whole of Personality. Freedom is wider than the will; the spontaneity of consciousness itself, and of the mind in its various constructive or creative activities, shows that Freedom is not limited to the will, but characterises also other forms of personal activity. In fact Freedom is not an attribute of mere parts or aspects but of the whole, and therefore of Personality considered as a whole.

Most important of all, we have to point out that Freedom, like Personality itself, admits of degrees in its personal manifestations. We saw earlier in this chapter that Personality, at the present stage of its history, is not yet fully developed; that it is imperfect as a whole even in the highest individuals, and that it varies in degree and intensity in all individuals. The power of perfect self-direction, assimilation and self-orientation which would distinguish a perfect personal whole is only imperfectly realised in individual cases; and in the same way there is a corresponding failure to realise the perfect ideal of Freedom.

Now in proportion as the Personality fails to achieve the character of a perfect whole, in the same proportion it is merely mechanical in its action, and therefore in the same proportion it becomes externally determined or un-free in its actions. The result is that the Personality is partly (so far as it is a whole) free, and partly bound or externally determined—that is to say, in so far as it is or behaves like a mechanism. Thus the fuller and more complete a Personality is the greater its power of central self-control, or the fuller its freedom. Weak characters have much less freedom than strong characters.

Temptation to the strong Personality finds itself enmeshed in the transforming power of a great system of central control which will actually turn it into a stimulus to the higher life; while the same temptation operating on a weak Personality finds little to withstand its force, and the resultant moral lapse is almost a mechanical equivalent of the temptation. Freedom is characteristic of the Whole just as Necessity is characteristic of Mechanism; and this is as true in regard to the moral action of the human agent as in abstract theory.

In what sense is the human agent free? In the everlasting controversy as to the freedom of the will, it has never been really denied that the will determines actions; that I can will to do this or that and do it accordingly. But Necessitarians and Determinists have contended that this will is itself not free, but determined by motives and conditions like all other natural events; that it is itself a mere link in the causal mechanical chain; and that the consciousness of freedom is really an illusion. Supporters of the Free Will theory have, on the other hand, contended that volitions are free, that the will in deciding on any course of conduct may act irrespective of motives or external conditions operating in it; and that this indeterminism is borne out by our consciousness of freedom of choice between various alternatives. Against this view there is not only the scientific evidence, but also the feeling that Freedom in this sense of unmotivated decision would be an exceptional capricious element in the orderly procedure of the universe. Capricious individual behaviour seems unworthy of such a world, and would certainly not accord with Holism such as we see it in the course of cosmic Evolution. In trying to arrive at the correct view, we must on the one hand discard mere physical determinism as being purely mechanical and in conflict with Holism in its organic and personal forms; and on the other we must recognise the universal orderly character of Holism, which does not admit of particular or individual caprice. And in this way we arrive at the idea of holistic, as distinguished from physical or mechanical, determinism. The Whole, and Personality in so far as it is a whole expressive of the individuating activity of Holism, are not and cannot be mechanically determined; they are self-determined in their characters as wholes. In other words, theirs is holistic as distinguished from mechanical determination. Freedom, not in the sense of individual caprice of choice, but in the sense of self-determination of a whole, or holistic determinism, is an inherent character of Personality, and flows from the very nature of Holism. In so far, however, as any human being is deficient in Personality his actions also tend to be a mechanical reflex of impulses and external conditions, and to that extent to lose the character of true freedom.

It is clear from the foregoing that Freedom is not merely a concept but becomes an ethical and personal ideal. Freedom is the full measure of self-realisation which each human being by its nature aspires to. It is not yet a firm possession of Personality. No doubt all Personality has it in some degree, just as every organism has it in a lower, more primitive form. But the freedom of a Personality is the measure of its development and self-realisation. It is the active power which secures the imperial legislative authority of the Personality, not only over its own rebellious impulses and tendencies, but even over the fleeting evanescent forms of thought and experience. In the ideal Personality Libertas and Imperium are identical. It is, in fact, the supreme prize to be contended for in the striving of each human being; and the extent of its inward realisation denotes the measure of the victory attained. To be a free Personality represents the highest achievement of which any human being is capable. The Whole is free; and to realise wholeness or freedom (they are correlative expressions) in the smaller whole of individual life not only represents the highest of which the individual is capable, but expresses also what is at once the deepest and the highest in the universal movement of Holism.

So much in regard to Freedom as the form and measure of personal development.

The problem of Purity is at bottom identical with that of Freedom; they are both but aspects of Wholeness. But while Freedom concerns the power of the Personality and means strength as against weakness, Purity means the harmony of the Personality through the elimination of alien elements and the co-ordination of all the personal tendencies in one harmonious whole of the spirit. A pure, free, homogeneous spirit is the ideal of Personality.

So long as disharmonies exist in the Personality and conflicts arise between different tendencies in it, so long the Personality will fall below its ideal of a pure homogeneous Whole. That ideal will be attained only when in the progress of personal development harmony and internal peace have been secured. It must not be supposed that the only manner in which this peace is possible is by the elimination or absorption of all the lower or earlier phases of personal evolution and the survival of the later higher phases. The Ideal Man will not be devoid of those passions and emotions which ordinarily war against the higher tendencies and aspirations of the Personality. But in the Ideal Man they will not cause conflict by contending for a dominating position in the Personality; they will be relegated to the subordinate position to which their more primitive crude character entitles them. In the Ideal Man the discords of ethical life will be composed, because there will be a harmonious correlation of higher and lower; the harmony will be the richer in proportion to the variety of elements which have been conserved and will thus combine to produce it. It takes all sorts to make the little world of Personality. The unity of character which the holistic movement aims at does not involve the destruction of the lower by the higher ethical factors, but the clear undisputed hegemony of the latter over the former, and the reduction of the former to a subordinate or servile position in the whole. It is this combination, in a harmonious form, of all grades of ethical evolution in the ideal Personality which will make it truly human, while at the same time it will be expressive of the universal order. To secure that harmony ought to be the supreme aim of the ethical individual.

From these remarks it will be clear how important a part moral discipline plays in the furtherance of the evolutionary holistic scheme. The life of the moral individual does not drift smoothly on like that of the happy gods, but is a constant gymnastic effort to strengthen the higher and to secure its dominance over the lower tendencies. The spiritual sublimation of the lower into the higher becomes the constant unremitting effort. The mechanical operation of Natural Selection is supplemented on the ethical plane by the conscious co-operation of those powers and agencies which have been evolved in the higher evolutionary processes. The contest is no longer left to be carried on by the blind activity of natural forces and animal instincts; but reason and conscience take a deliberate hand in the great issue of Holism. The progress of Holism involves that mere Naturalism shall be superseded or at least sublimated at the higher stages of evolutionary progress into the deeper holistic factors which have appeared on the scene in Personality. And the object of this conscious moral discipline should not be the ascetic suppression of primitive healthy human instincts, but their refinement and sublimation, their subordination and co-ordination in the growing whole of the Personality under the hegemony of the later and higher ethical factors.

While moral discipline thus plays an important part in personal evolution, it must not, however, be supposed that Personality should go on for ever oppressed by an overpowering sense of Duty, and should hear for ever the thundering reverberations of the Categorical Imperative. No doubt when the person at his moral awakening or some other moral crisis in his life, first hears the trumpet-call of Duty, the effect is tremendous. But the thunder should die away into the still small voice of the inner life; the apparently alien forbidding aspect of Duty should be assimilated into the quiet normal impulses of the Personality; moral discipline should so thoroughly become second nature to the ethical warrior that its effects will be there without its operation being felt. The Personality should reach such a standard of purity and homogeneity that there will be no alien stuff in it to offer resistance to the promptings of Conscience or Duty or to cause friction or disquietude in the soul. The highly developed and disciplined Personality, pure and homogeneous in itself, and in harmony with universal Holism, and thus finely responsive to all things true and good and fair in the universe, will not only embody the ancient Greek ideal of $\sigma\omega\phi\rho\sigma\sigma\acute{u}\mu\eta$, or moderation and self-control, but will also come to realise both the Stoic and the Epicurean ideal of $\dot{a}\tau a\rho a\xi\acute{\iota}a$, or tranquillity of soul, and finally to know that peace of God, passing all understanding, which is the supreme promise of the Buddhist no less than of the Christian religion.

The ethical message of Holism to man is summed up in two words: Freedom and Purity. And from what we have just seen it is clear that these two grand ethical ideals are at bottom identical. The function of the ideal of Freedom is to secure the inward self-determination of the Personality, its riddance of all alien obstructive elements, and thus its perfection as a pure, radiant, transparent, homogeneous self-activity. In other words, the function of Freedom is to attain Purity in the Personality. And similarly the function of the ideal of Purity is to afford free play to the inward self-determination and self-activity of the Personality by removing all external impediments, all stains and impurities, all vice, cowardice, intemperance and injustice, all evil and ugliness; in short, all elements alien to the nature of the Personality, and thus to realise the Ideal of Freedom in the Personality.

This statement differs considerably from the usual ways of formulating the *Summum Bonum* or Ethical End. The Pleasure of the Hedonist, the Good of the Intuitionist, and all the other abstract formulations of the Ethical End appear partial and one-sided from the holistic point of view. The end of Personality does not lie outside it but is given inwardly. As Goethe has so well said of Life: "Der Zweck des Lebens ist das Leben selbst." Even more truly one may say that the Whole knows no end beyond or outside itself. The object of the holistic movement is simply the Whole, the Self-realisation and perfection of the Whole. And the same is true of Personality in so far as it is a whole. Its object is to achieve self-realisation, to realise its wholeness, to attain freedom not in a selfish, egoistic sense but in the universal holistic order. Holistic self-realisation is no doubt pleasurable to the individual; but the pleasure is a mere side issue and by-product, so to say, of the striving towards wholeness in the individual life and character, and the same may be said in regard to all the other particular ends and aims usually considered worthy of our serious endeavour. Learn to be yourself with perfect honesty, integrity and sincerity; let universal Holism realise its highest in you as a free whole of Personality; and all the rest will be added unto you—peace, joy, blessedness, happiness, goodness and all the other prizes of life. Nay, more: the great evils of life—pain, and suffering, and sorrow—will only in the end serve to accelerate the holistic progress of the Personality, will be assimilated and transformed in the spiritual alchemy of the Personality and will feed the flame of the pure and free soul.

It would be a mistake to look upon the ideal of personal holistic self-realisation as merely egoistic. No doubt in some cases the subjective selfish features may predominate; but earnest men will always find that to gain their life they must lose it; that not in self but in the whole (including the self) lies the only upward road to the sunlit summits. We mostly move in the channels worn by social usage or convention and are influenced by personal and social impulses such as ambition, patriotism, love of money or power. But Holism is deeper than any of these. The inner call of Holism is to none of these things in themselves and for their own sake, but to its own victory in the personal life; to unity, freedom and free plastic power for the Personality; to active moral efficiency and the suppression of harmful elements in the personal life: in a word, to the wholeness and perfection of the Personality. The response to that call in the personal life constitutes the great inner drama, the warfare in the Soul, which issues either in the attainment of Wholeness and Freedom and membership in the immortal Order of the Whole, or otherwise in defeat, enslavement and death.

CHAPTER XII

THE HOLISTIC UNIVERSE

Summary.—The fundamental, seminal character of the concept of Holism is bound to affect our general views of the nature of the universe, our *Weltanschauung*, and this chapter deals with this wider aspect of Holism.

Holism has been presented in the foregoing chapters as the ultimate synthetic, ordering, organising, regulative activity in the universe which accounts for all the structural groupings and syntheses in it, from the atom and the physico-chemical structures, though the cell and organisms, through Mind in animals, to Personality in man. The all-pervading and ever-increasing character of synthetic unity or wholeness in these structures leads to the concept of Holism as the fundamental activity underlying and co-ordinating all others, and to the view of the universe as a Holistic Universe.

On a strict and narrow view Science may consider the concept of Holism as extra-scientific, as giving a metaphysical and not a scientific explanation of things. But this would be a mistake for three reasons. In the first place, the conclusion to which Science is pointing, namely, that the whole universe, inorganic as well as organic, is the expression of cosmic Evolution, necessitates a ground-plan which will formulate and explain this vast scientific scheme of things. Mere pre-occupation with detailed mechanisms will no longer suit the immensely enlarged scope of present-day Science. In the second place, Science has already had to assume such ultra-scientific entities as, for instance, the ether of space, as necessary to give a coherent explanation even of purely physical phenomena. And the correlation of the physical, and organic, and psychical in one vast scheme of Evolution similarly necessitates much more widely operative factors than have been hitherto recognized. Holism is far more necessary for cosmic Evolution than was the ether for light transmission. In the third place, Holism is essentially no more ultra-scientific than are life and mind; it is simply a wider concept than either and is the genus of which they are the species. And it enables all the evolutionary phenomena of Nature to be co-ordinated under and traced to the same operative factor.

The New Physics has traced the physical universe to Action; and Relativity has led to the concept of Space–Time as the medium for this Action. Space–Time means structure in the widest sense, and thus the universe as we know it starts as structural Action; Action which is, however, not confined to its structures, but continually overflows into their "fields" and becomes the basis for the active dynamic Evolution which creatively shapes the universe. The "creativeness" of evolutionary Holism and its procedure by way of small increments or instalments of "creation" are its most fundamental characters, from which all the particular forms and characteristics of the universe flow.

The ignorance or neglect of these two fundamental characters accounts for the elements of error involved in certain widely held world-conceptions, such as Naturalism, Idealism, Monadism and Spiritual Pluralism or Panpsychism. Naturalism is wrong where it fails to recognise that there is creative Evolution, and that real new entities have arisen in the universe, in addition to the physical conditions of the beginning. Idealism is wrong where it fails to recognise that the Spirit or Psyche, although now a real factor, did not exist either explicitly or implicitly at the beginning, and has arisen creatively in the course of organic Evolution. The Monadism of Leibniz and his modern sympathisers, while a great advance in that it recognises the inward holistic element in things and persons, yet goes wrong when it attributes an element of Mind or Spirit to physical things like atoms or chemical structures. While things are wholes they are not yet souls; and the view of the universe as a Society of Spirits ignores the fact that spirit is a more recent creative arrival in the universe and cannot be retrospectively antedated to the earlier material phase. Spiritual Pluralism is a modern refinement of Monadism and similarly subject to the criticism that it fails to recognize the really creative character of Evolution.

This is a universe of whole-making, not of soul-making merely. The view of the universe as purely spiritual, as transparent to the Spirit, fails to account for its dark opaque character ethically and rationally; for its accidental and contributory features, its elements of error, sin and suffering, which will not be conjured away by an essentially poetic world-view. Holism explains both the realism and the idealism at the heart of things, and is therefore a more accurate description of reality than any of these more or less partial and one-sided world-views.

Nature or the Universe is sometimes metaphorically spoken of as a Whole or The Whole. Sometimes it is even personified, and the trend of Evolution then becomes the Purpose of some transcendent Mind. All this is, however, unwarranted by the facts and unnecessary as an explanation of Evolution. Holism as an inner evolving principle of direction and control in all Evolution is enough; it underlies the variations which arise and survive in the right direction, and it creates in the "field" of Nature a general environment of internal and external control. The "wholeness" or holistic character of Nature appears mostly in this field or environment of Nature, with its friendly intimate influences, and its subtle appeal to all the wholes in Nature, and especially to the spiritual in us. The fact is that the Holism in Nature is very close to us and a real support in all our striving towards betterment. Our aspiration is its inspiration, and it is thus the inner guarantee of eventual victory in spite of all set-backs and defeats.

THIS is not a treatise on Philosophy; not even on the philosophy of Nature; not even on the philosophy of Evolution. It is an exploration of one idea, an attempt to sketch in large and mostly vague, tentative outline the meaning and the consequences of one particular idea. But that is a seminal idea; indeed it is here presented as more than an idea, as a fundamental principle operative in the universe. As such it is bound to affect our general view of the nature of the universe. I therefore come in this concluding chapter to consider what Holism means for our general world-view, our *Weltanschauung,* and as briefly as possible to sum up the bearing which the argument of the preceding chapters must have on such a general conception of the universe.

Holism has been our theme—Holism as an operative factor in the universe, the basic concept and categories of action of which can be more or less definitely formulated. I have in the broadest outline sketched the progress of Holism from its simple mechanical inorganic beginnings to its culmination in the human Personality. All through we have seen it at work as the fundamental synthetic, ordering, organising, regulating activity in the universe, operating according to categories which, while essentially the same everywhere, assume ever more closely unified and synthetic forms in the progressive course of its operation. Appearing at first as the chemical affinities, attractions and repulsions, and selective groupings which lie at the base of all material aggregations, it has accounted for the constitution of the atom, and for the structural organising of atoms and molecules in the constitution of matter. Next, after some gaps which are being energetically explored by biology and bio-chemistry, and still operating as a fundamental synthetic selective activity, it has emerged on a much higher level of organisation in the cell of life, and has again been responsible for the ordered grouping of cells in the life-structures of organisms, both of the plant and the animal type, and in the progressive complexifying of these structures in the course of organic Evolution. The synthetic activity in these organic structures has been so far-reaching that the independent existence of the original unit cells has sometimes been questioned, and the organism has been taken as the synthetic unit, of which the cell is but a defined portion of nucleated protoplasm.[1] In other words, the organic synthesis of cells has been such as practically to lead to the suppression of the individual cells as such. Next, in the higher animals and especially in man, Holism has emerged in the new mutation or series of mutations of Mind, in which its synthetic co-ordinating activity has risen to an unheard-of level, has turned in upon itself and become experience, and has achieved virtual independence in the form of consciousness. Finally, it has organised all its previous structures, including mind, in a supreme structural unity in Human Personality, which has assumed a dominating position over all the other structures and strata of existence, and has in a sense become a new centre and arbiter of reality. Thus the four great series in reality—matter, life, mind and Personality—apparently so far removed from each other, are seen to be but steps in the progressive evolution of one and the same fundamental factor, whose pathway is the universe within us and around us. Holism constitutes them all, connects them all and, so far as explanations are at all possible, explains and accounts for them all. Holism is matter and energy at one stage; it is organism and life at another stage; and it is mind and Personality at its latest stage. And all its protean forms can in

[1] Doncaster: *Introduction to Study of Cytology*, pp. 3–4.

THE HOLISTIC UNIVERSE

a measure be explained in terms of its fundamental characters and activities, as I have tried to show. All the problems of the universe, not only those of matter and life, but also and especially those of mind and personality, which determine human nature and destiny, can in the last resort only be resolved—in so far as they are at all humanly soluble—by reference to the fundamental concept of Holism. For this reason I have called our universe "the Holistic universe," as Holism is basic to its constitution, its multitudinous forms and its processes, its history in the past, and its promise and potency for the future.[1]

The scientist, viewing my claims for Holism in the dry light of Science, might perhaps feel tempted to demur to them. He might object that Holism is a mere assumption which may have a philosophical or metaphysical value, but that it has no scientific importance, as it cannot be brought to the test of actual facts and experiments. Holism as here presented, he will say, is not a matter for Science; it is an ultra-scientific entity or concept. It falls outside the scope

---

[1] Professor Lloyd Morgan has made the creative or emergent character of Evolution the theme of his book on "Emergent Evolution," and it has been suggested to me that I should explain my relation to it. The fact is that my views had a different origin from his, and that they had been matured and the whole of this book written before I saw his interesting and suggestive volume. The result is that, in spite of many surprising similarities of thought, there remains an essential diversity in our themes as well as in our emphasis even on those matters on which we apparently agree. To him emergence of the new in the evolution of the universe is the essential fact; to me there is something more fundamental—the character of the wholeness, the tendency to wholes, ever more intensive and effective wholes, which is basic to the universe, and of which emergence or creativeness is but one feature, however important it is in other respects. Hence he lays all the emphasis on the feature of emergence, while I stress wholes or Holism as the real factor, from which emergence and all the rest follow. To me the holistic aspect of the universe is fundamental, and appears to be the key position both for the science and for the philosophy of the future.

Besides, Professor Lloyd Morgan makes the psychical factor the correlate at all stages of the physical factor, thus in effect getting back to the Spinozist position that all bodies, even inorganic matter, are *animata* in their several degrees. This view seems to be a reversion to the preformation type of Evolution and to be destructive of all real effective "emergence." In any case it is wholly different from the view of creative advance consistently put forward in this book.

of Science, and the explanation of things which it purports to give is not a scientific explanation. Even assuming that there is such an activity as Holism at work in the universe, it would have no value for Science. To be of interest to Science, it must make a difference to actual facts and therefore be capable of experimental verification. But clearly Holism, owing to its pervasiveness and universality, cannot be so tested. As its presence would not be revealed by an examination of the particular facts, mechanisms and phenomena with which Science deals, it is unnecessary for Science to take any further interest in it.

I hope I have fairly summarised the attitude which Science might perhaps feel impelled to adopt towards the claims I have put forward on behalf of Holism. And I would reply by pointing out what seems to me to be the weakness or rather the one-sidedness and partialness in this strictly scientific attitude. Science seems to me to take too narrow a view of her sphere and functions when she confines herself merely to details, to the investigation and description of the detailed mechanisms and processes in regard to matters falling within her province. A description of analytical details, however true so far as it goes, is not yet a full and proper account of the thing or matter to be described. It is not enough; the details must be supplemented by a description which will take us back to the whole embracing those details. The anatomy and physiology of a plant would surely not be sufficient as a description of the plant itself. No description of the parts is a complete description of the whole object; it is only a partial description, and falls short of a true and full account in proportion as the object partakes of the character of a whole; where the object, for instance, is what I have called a biological or psychical whole. We may say generally that wherever an object shows structure or organisation (as every object does) a full description of it would involve at the very least an account of this structure or organisation as a whole, in addition to its detailed mechanisms and functions. And where many objects show similar or related structures, a proper description would involve an account of the ground-plan of organisation affecting them all. Thus in regard to organic and inorganic Evolution, where the whole world of matter and life and mind can be grouped into progressive series of structures from the beginning to the end, a scientific account of the universe would necessarily involve the working out of the universal ground-plan which expresses this Evolution. And it can but add to the value of such a ground-plan that it is not merely descriptive but also attempts to be self-explanatory. A plan or scheme is by its very nature not properly stated unless it is not merely described but also explained and accounted for as far as possible. Now I ask, what else is Holism but such an attempted ground-plan of the universe, which is of a self-explanatory character, a ground-plan which makes the whole scheme the progressive operation and effect of a given cause? It may be objected that ultimate causes lie beyond the purview of Science. But even so the descriptive ground-plan of Holism would remain and would challenge serious consideration on scientific grounds. To me the issue seems quite simple. So long as Science eschewed all wider viewpoints (as she modestly did in her earlier years) and confined her attention to particular areas of facts, such as are embraced by the separate sciences, she was quite entitled to look upon a general explanatory ground-plan of Evolution as too ambitious for her and as falling outside her proper sphere. But once she abandons this sectional standpoint and comes to look upon the entire universe as evolutionary (as she now does), she is bound to examine a scheme such as is here put forward on its merits as falling within her universal province.

Science has been compelled in other instances to complete and support her account of detailed processes by the assumption of factors which lie beyond the area of observation, but without which the detailed processes become unintelligible. Thus the assumption of the ether of space was resorted to as the basis of the undulatory theory for the transmission of radiant energy. Although ether admittedly lies beyond the area of scientific observation and experiment, and no test however delicate has ever revealed its actual existence, it was long accepted as one of the conceptual entities which were necessary to complete the coherent system of Science, and indeed as a real physical element in the universe. It is true that ether seems to have fallen on evil days and that its existence or conceptual necessity is being more and more questioned by various groups of physicists. But it has admirably served its purpose as a scientific hypothesis, and the legitimacy of such a hypothesis was never questioned by even the most rigid school of scientists. And I would submit that the case for Holism is much stronger than that for ether ever was, as ether was meant to account only for one particular group of phenomena in physics, while Holism in the main phases of its development is necessary to account for the facts and phenomena of Evolution, both organic and inorganic. The plain fact is that, as our intellectual outlook widens and the intellectual horizons recede more and more, the domain of Science is undergoing an ever greater expansion, and therefore the formulation of new principles and new concepts embodying them becomes necessary for the support and the coherence of the whole vast scheme of Science. Science is thus for ever encroaching on the domain of philosophy and the other great disciplines of the Reason or the Spirit, and it becomes ever more difficult to confine her activities within the old orthodox limits. Holism no doubt breaks new ground; it is here intended as the basis of a new *Weltanschauung* within the general framework of Science; it is meant to be the foundation of a new system of unity and inward character in our outlook upon the universe as a whole. But it does not fall outside the province of Science in the larger sense. And it does not introduce strange, alien concepts into the sphere of Science.

I would also point out that the scientific objection to Holism as above formulated would, in fact, be identical with the objection which mechanistic Science has taken to life and mind as operative factors in the universe—an objection which Science is feeling herself ever more strongly compelled to overrule. The difficulty to verify Holism in the detailed mechanisms and functions would be the same as the difficulty to verify life and mind as operative factors in organic and human structures. Holism is really no more than an attempt to extend the system of life and mind, with the necessary modifications and qualifications, to inorganic Evolution, and to show the underlying identity of this system at all the stages of Evolution. In life the character of the system becomes clear, in mind still more so. That is no reason to look upon it as non-existent in the case of matter. The facts submitted in the foregoing chapters disclose a more or less connected, graduated, evolutionary series covering the phenomena of matter no less than of life and mind. Holism is a concept and a factor which formulates and accounts for the fundamental groundplan of this series. It is therefore very much of the same order of ideas as life and mind, and stands or falls very much by the same lines of reasoning as they.

The graduated serial character of the universe has led to the theory of Evolution. But it is clear that that serial character opens up still greater questions of sources and origins. A connected graduated system of facts implies not only a particular method of their becoming, such as the theory of Evolution formulates, but also a common origin and a common propelling force or activity behind the system. In life and still more in mind we get clear indications of this origin and this activity. All that remains is to take a wider view and to bring all the facts and phenomena of the universe within the scope of this common method and origin. We then reach the concept of Holism as embracing life and mind, but covering a much wider area and forming, in fact, the genus of which they are the species. All Evolution then becomes the manifestation of a specific fundamental, universal activity.

Having thus attempted to vindicate Holism as a proper scientific concept in the wider sense, let us now proceed to sketch the main distinguishing features of the Holistic universe, in other words, of the conception of the universe which results from the principles discussed in the foregoing chapters. The final net result is that this is a whole-making universe, that it is the fundamental character of this universe to be active in the production of wholes, of ever more complete and advanced wholes, and that the Evolution of the universe, inorganic and organic, is nothing but the record of this whole-making activity in its progressive development. Let me briefly summarise the main points in the preceding argument which lead up to this result.

We have seen that this is an essentially and wholly active universe; that its apparent passiveness as matter is nothing but massed energy, and that activity is therefore its fundamental character. Indeed, energy itself is too narrow and metrical a term to do justice to this character of the physical universe as concrete activity. Activity in time, energy multiplied by time, *Action* as it is technically called in physics, is the physical basis of the universe as a whole, and nothing besides. The universe is a flowing stream in Space–Time, and its reality is not intelligible apart from this concept of activity. So much the new Relativity has made us realise; and to this conclusion the profoundest reflections on the nature of the universe also tend. For us, constituted as we are, the universe starts and takes its origin in Action. With deeper meaning than ever before we realise that " Im Anfang war die That." It is, of course, conceivable that much lies beyond and back of this beginning as it appears to us. It may be that the universe of Action has itself evolved out of a prior order which lies beyond human ken; that there is an infinite regress of celestial Evolution into time past; and that the physical universe as it now appears to or is conceived by us is the evolved result of inconceivable prior developments. We do not know, and speculation would be barren and futile.

The physical stuff of the universe is therefore really and truly Action and nothing else. But when we say that, when we make activity instead of matter the stuff or material of the universe, a new view-point is subtly introduced. For the associations of matter are different from those of Action, and the dethronement of matter in our fundamental physical conception of the universe and its replacement by Action must profoundly modify our general outlook and viewpoints. The New Physics has proved a solvent for some of the most ancient and hardest concepts of traditional human experience and has brought a *rapprochement* and reconciliation between the material and organic or psychical orders within measurable distance. I must refer to the concluding portion of the third chapter for a statement of this far-reaching advance which has been made by Science within this century. That is the contribution of the New Physics to the new outlook.

Action in Space–Time is necessarily structural; indeed Space–Time supplies the co-ordinates, the framework of the Activity which is the ultimate stuff of the world. Space–Time is the structure; hence Action in Space–Time, in the first phase of Holism, is purely structural and mechanistic, as we saw in Chapter VII. The recognition of the fundamental structural character not only of matter but of the whole universe is the contribution of the Relativity theory to the new outlook. The physical world thus becomes at bottom structural Action, Activity structuralised in bodies, things, events. Thus arises the apparent material universe which surrounds us and in our bodies forms part of us.

What is the next step? Action does not come to a stop in its structures, it remains Action, it remains in action. In other words, there is more in bodies, things and events than is contained in their structures or material forms. All things overflow their own structural limits, the inner Action transcends the outer structure, and there is thus a trend in things beyond themselves. This inner trend in things springs from their very essence as localised, imprisoned Action. From this follow two important conclusions. The first is the concept of things as more than their apparent structures, and their " fields " as complementary to their full operation and understanding. A thing does not come to a stop at its boundaries or bounding surfaces. It is overflowing Action, it passes beyond its bounds, and its surrounding "field" is therefore essential not only to its correct appreciation as a thing, but also to a correct understanding of things in general, and especially of the ways in which they affect each other. I have tried, at various points in the foregoing discussion, to emphasise the great importance of this concept of "fields" not only for physics but also for biology and philosophy.

The second and more important conclusion is the great fact and concept of Evolution. The inner character of the universe as Action expresses itself in actuality as a passage, a process, a passing beyond existing forms and structures, and thus the way is opened up for Evolution. The actual character of Evolution can, of course, be concluded and known only from the facts, and is not a matter of logical inference deducible from the nature of the universe as Action. But if activity is the essence of the universe we see more easily why the universe is evolutionary and historical rather than static and unchangeable. There is a passage, a process, a progress, but its characters can only be determined by a study of the facts of the passage. It may turn out to be merely a movement of combinations and groupings; or it may turn out to be an unfolding, explication and filling out, an *evolutio* in the stricter sense; or it may turn out to be a real creative Evolution such as we have seen it to be. Real Evolution requires other concepts besides those of Action and structure; and these concepts can only be derived from experience. Thus the creativeness of Evolution is a conclusion not so much from theory as from the empirical facts. And the exact nature of this creativeness is unknown in some respects and remains a problem for the future to solve. A still wider survey and closer scrutiny of the facts lead to the conception of Holism which accounts not only for the structural combinations of bodies, things and events, but for all the progressive series of unities and syntheses which have arisen in the cosmic process.

Assuming that Holism and the nature of wholes in the universe have been sufficiently explained in the foregoing chapters, I now proceed to compare the world-view to which Holism leads with those which resemble or touch it at various points and yet are essentially different from it.

The Holistic view agrees with the Naturalistic conception of physical science in giving the fullest importance to the physical aspect of the universe. It does full justice to the structural and mechanistic characters of Nature, and indeed it considers Mechanism simply an earlier phase of Holism, and therefore perfectly legitimate up to a point. It affirms the validity of the fundamental laws and principles of physics not only for inorganic bodies but also for organisms, in so far as they are material. It represents the organic order as arising from and inside the inorganic or physical order without in any way derogating from it. If in the end it erects on the physical a superstructure which is more and more ideal and spiritual, that does not mean a denial of the physical. The idealism of Holism does not deny matter, but affirms and welcomes and affectionately embraces it. If Holism begins as realism and ends as idealism, it does not spurn or deny its own past; in Holism both realism and idealism have their proper place and function and indeed find their justification and reconciliation. It breaks with Naturalism only at the point where Naturalism becomes purely materialistic, and in effect denies the creative plasticity of Nature, presents Nature as an anatomical museum, as a collection of dead and dried *disjecta membra*, instead of the interwoven body of living, creative, progressive unities and syntheses which she essentially is. Naturalism represents the universe as a vast reservoir of energy, unalterable in amount but steadily deteriorating in character, subject to immutable laws and fixed equations which prevent anything essentially new from ever arising or having arisen. It thus negatives the concept of creative Evolution except as a mere figure of speech. It presents life and mind as mere wandering insubstantial shadows on the shores of this ocean of energy; the great Mirage of Evolution broods over the waters; and Man himself, so far from being a creative factor in reality as a whole, becomes an impotent spectator of this melancholy scene, wrapped up in the illusions of his own self-consciousness. Such a view of the universe seems to me hopelessly one-sided and distorted, and comes into direct conflict with large and important bodies of facts of experience which cannot be denied or reasoned away. To me the rock on which Naturalism must split is the fact of creative Evolution. In the first chapter I pointed out that the old materialistic Naturalism is inconsistent with a clear and frank recognition of the great fact of Evolution, and in the body of this work I have tried to drive the point further home.

Creativeness is the key-word and the key-position, not only as far as Naturalism is concerned, but also as regards those other world-conceptions which are most hostile to Naturalism, such as the various modern forms of Spiritual Idealism. Naturalism imposes the past on the present and the future; Idealism, again, imposes the present and the future on the past. Both implicitly deny that creative Evolution which shows the universe historically as a gradual transformation, as a real creative process moving from the real structures of the past to the real structures of the future; and therefore as a system which in its historical development embraces and gives justification to both contrasted points of view. To view the ideal or spiritual element in the universe as the dominant factor is to ignore the fact that the universe was before ever the ideal or spiritual had appeared on the horizon; that the ideal or spiritual is a new and indeed recent creation in the order of the universe, that it was not implicit in the beginnings and has not been reached by a process of unfolding; but that from a real pre-existing order of things it has been creatively evolved as a new factor; and that its importance to-day should not be retrospectively antedated to a time when the world existed without it. Where was the Spirit when the warm Silurian seas covered the face of the earth, and the lower types of fishes and marine creatures still formed the crest of the evolutionary wave? Or going still further back, where was the Spirit when in the Pre-Cambrian system of the globe the first convulsive movements threw up the early mountains which have now entirely disappeared from the face of the earth, and when the living forms, if any, were of so low a type that none have been deciphered yet in the geological record? Where was the Spirit when the Solar System itself was still a diffuse fiery nebula? The evolutionary facts of Science are beyond dispute, and they support the view of the earth as existing millions of years before ever the psychical or spiritual order had arisen; and what is true of the earth may be similarly true of the universe as a whole. The fact that we have to grasp firmly in connection with creative Evolution is that, while the spiritual or psychical factor is a real element in the universe, it is a comparatively recent arrival in the evolutionary order of things; that the universe existed untold millions of years before its arrival; and that it is just as wrong for Idealism to deny the world before the appearance of Spirit, as it is for Naturalism to deny Spirit when eventually it did appear in the world.

Creative Evolution seems to move forward by small steps or instalments or increments of creativeness. Why there should be this discontinuity rather than a smooth continuous advance we cannot say; we can but note the fact, which seems to be a universal phenomenon. Not only does matter in its atomic and elemental structure show this minute discontinuity, but the electric elements in the atom, and in the electric current generally, and the quanta of heat and radiant energy show the same remarkable feature. Thus the unit character of Action and Structure is reproduced in the unit character of Evolution and of nuclear change in the cell. There is real creation as distinct from mere combinations of pre-existing units or mere unfolding of implicit elements; but this creation is not consummated in one supreme creative Act; nor is it evenly and uniformly distributed throughout all time. Its distribution is unevenly spread in minute parcels over the whole almost infinite range of Evolution. Evolution thus becomes a long-drawn-out process of creation, in which the new for ever arises by slow and minute increments from the old, or rather by way of the old, as it is not known how the new actually arises from the old. As I have explained in Chapter VII, Holism is the presiding genius of this advance. It determines the direction of the advance, and it incorporates the new element of advance synthetically with the pre-existing structure. It thus harmonises the old and the new in its own unity; it synthesises Variation and Heredity; and by slow degrees and over enormous periods of time carries forward the creative process from the most simple, primitive, inorganic beginnings to the most exalted spiritual creations. From the atom to the Soul, from matter to Personality is a long way, marked by innumerable steps, each of which involved a real creative advance and added something essentially new to what had gone before. Such seems to be the nature of Evolution, and it appears to be fatal alike to the retrospective interpretation of the universe according to Idealism, and the prospective interpretation according to Naturalism. Mind or Spirit did not exist at the beginning, either implicitly or explicitly; but it does most certainly exist now as a real factor.

Another world-conception which may be considered as having considerable affinities with the Holistic view is that of Leibniz's Monadology. The resemblance is, however, confined to certain aspects of the respective central ideas; beyond those aspects the two views are totally and essentially different. There is a close resemblance between the central ideas of wholes and monads; that is all. The unities and units which exist in Nature seemed to Leibniz to be of the greatest importance for the interpretation of the universe; not the One but the Many and their intimate nature seemed to him to supply the key to the great riddle. I have in the foregoing reached the concept of wholes by a different process of reasoning from that followed by Leibniz, but the result looks very much like that arrived at by him along different lines. And the convergence of the two views from totally different standpoints would appear to suggest that there is a substantial element of truth and value in the concept of wholes, as there undoubtedly is in the Leibnizian theory of Monads. They agree in having an innerness, in being little worlds of their own, with their own inner laws of development and with a certain measure of inner self-direction or self-conservation which makes them partial mirrors or expressions of the greater reality. But the monads according to Leibniz are essentially spiritual entities or selves, conceived on the analogy of the human mind, and their activities are of a purely psychical character such as perception. They are, moreover, absolutely closed, isolated, self-contained units, each with its own immutable inner system, uninfluenced by any other monad; and all maintained in harmony with each other by some divine pre-established order outside of them. The greater and lesser selves of the universe lead their own inner self-existences, without any contact between one another, and only the divine interposition maintains a Pre-established Harmony between them. There is a scale of these monads, from the lowest most simple, such as atoms or molecules, whose confused perceptions produce the world of matter, to the highest most complex in the universe, such as human minds, whose clear and distinct perceptions produce the world of spirit. God himself is but the Supreme Monad of monads on this view. It will be seen how different this monadic conception is from that of wholes developed in this work. In the first place, wholes are not all spiritual entities, and the world is not a hierarchy of spirits exclusively, as Leibniz conceived it. Spiritual wholes are merely the apex and crowning feature of the universe, while non-spiritual (material or organic) wholes compose its earlier phases. In the second place, wholes are not closed, isolated systems externally; they have their fields in which they intermingle and influence each other. The Holistic universe is a profoundly reticulated system of interactions and inter-connections rising into a real society in its later phases. In the third place, genetic relationships connect the entire Holistic universe. Wholes from the lowest to the highest are akin and form one great family, and are derived from one another in the process called Evolution. In the fourth place, it is the ascertainable character of this evolutionary process which holds all the wholes together in one vast network of adaptations and harmonious co-ordinations, and not some mystic assumed Pre-established Harmony. Leibniz, while he correctly guessed the real secret in his idea of Monads, missed yet the true explanation through not having any knowledge of creative Evolution, such as the deeper science of our day has revealed to us. To him Evolution was a mere unfolding of an implicit content; he adhered to the traditional preformation views of his day as well as to the current belief in the fixity of species. He could not, therefore, realise the idea that monads were genetically related and evolved; and that the order which underlay the series of monads from the lowest to the highest was of a creative character. In the absence of genetic relationships and creative Evolution, he had to make shift with the notions of isolated inner selves and a pre-established harmony. We may therefore say that just as both Naturalism and Idealism are shattered on the rock of creative Evolution, so likewise the Monadology, however valuable and suggestive in other respects, founders on that same rock, which was, however, still secret and undisclosed to the science of Leibniz's time. But for that ignorance who knows whether Leibniz might not have elaborated a far more adequate and suggestive Holistic conception than that contained in this poor effort!

The astonishing thing is that thinkers of our own time, who are not only conversant with the idea of creative Evolution but convinced adherents of it, fail to adjust their view-points to it. Thus the late Professor James Ward, who advocated the view of Evolution as epigenesis or creative synthesis, and whose Pluralism has close affinities to the Monadology, seems yet to have failed to realise that his view of Evolution as creative was in conflict with his spiritual Pluralism or Panpsychism. His Pluralistic universe also consisted entirely of spiritual monads or entities, and this implied the possession of spiritual or psychical characters not only on the part of the higher monads, like persons, but also on the part of the most rudimentary monads, such as atoms and chemical compounds. Spinoza, who otherwise differed widely from Leibniz, had also assumed that all things were in their several degrees *animata*, but he had the excuse of being, like Leibniz, ignorant of the idea of creative Evolution. But Ward, in spite of his fuller knowledge, calmly follows Leibniz and Spinoza in their error. The plain fact, of course, is that psychism or spiritualism can by no stretch of language be ascribed to mere bits of matter or energy or physical entities like atoms or chemical compounds without the gravest confusion. The very idea of creative Evolution or epigenesis is that both life and mind are later creations in the evolutionary series, and cannot possibly be antedated to the mere physical level of Evolution. There is not a great Society of Spirits in the universe, of which Persons and Things, Souls and Atoms alike are members on the same spiritual footing. When the term " Spiritual " is stretched that far and spread that thin, it loses all real value and becomes a mere empty figure of speech. There is indeed no such spiritual Society of the whole universe, but there is the Holistic order, which is something far greater, and stretches from the beginning to the end, and through all grades and degrees of holistic self-fulfilment. Holism, not Spiritualism, is the key to the interpretation of the universe. Mind is not at the beginning but at the end, but Holism is everywhere and all in all. If the universe were a great spiritual Society of lower and higher souls or spirits, Evolution as creative would become meaningless; it would be merely a process of explication of the implicit spirituality (if any) inherent in the universe. The Holistic view thus not merely negatives the far-reaching spiritual assumptions of the Monadology, or Panpsychism, but it is also in firm agreement with the teachings of science and experience. Nor does it, in fact, detract from the value or importance of the universe. It but impresses on us the necessity of that great lesson of humility which is the ethical message of Evolution. It shows that values should not be confused with origins, and that from origins the most lowly may be raised values the most exalted and spiritual in the order of the universe. The Great Society of the universe leaves a place for the most humble inanimate inorganic structure no less than for the crowning glory of the great soul. To conceive the universe otherwise is to indulge in anthropomorphism, which may be pleasing to our vanity, but in reality detracts from the richness and variety of the universe. The Holistic universe embraces all the real structures from the lowest to the highest in their own right and as they are, without decking them in spiritual habiliments which are alien to their true nature. This world, in the noble language of Keats, is indeed the valley of soul-making; but it could not be that if the valley itself consists of nothing but souls. To those who have the deepest experience of life, this world is not only the upward path for the soul, but a very hard and flinty one. To attempt to pave that rugged way with the roses of the spiritual order would be a profound mistake from every point of view.

To say this is not to assume that there is anything alien or antagonistic between the human soul and the natural environment in which it finds itself in this world. There is not only poetic value but profound truth in the spiritual interpretation of Nature to which Wordsworth and other great poets of Nature have accustomed us. And that truth is not merely due to the creative part which mind plays in the shaping and fashioning of Nature. It is not merely that we invest Nature with our own emotional attributes. It is, in fact, to be traced to far deeper sources in our human origins. For we are indeed one with Nature; her genetic fibres run through all our being; our physical organs connect us with millions of years of her history; our minds are full of immemorial paths of pre-human experience. Our ear for music, our eye for art carry us back to the early beginnings of animal life on this globe. Press but a button in our brain, and the gaunt spectres of the dim forgotten past rise once more before us; the ghostly dreaded forms of the primeval Fear loom before us and we tremble all over with inexplicable fright. And then again some distant sound, some call of bird or smell of wild plants, or some sunrise or sunset glow in the distant clouds, some mixture of light and shade on the mountains, may suddenly throw an unearthly spell over the spirit, lead it forth from the deep chambers, and set it panting and wondering with inexpressible emotions. For the overwrought mind there is no peace like Nature's, for the wounded spirit there is no healing like hers. There are indeed times when human companionship becomes unbearable, and we fly to Nature for that silent sympathy and communion which she alone can give. Some of the deepest emotional experiences of my life have come to me on the many nights I have spent under the open African sky; and I am sure my case has not been singular in this respect. The intimate *rapport* with Nature is one of the most precious things in life. Nature is indeed very close to us; sometimes perhaps closer than hands and feet, of which in truth she is but the extension. The emotional appeal of Nature is tremendous, sometimes almost more than one can bear. But to explain it we need not make the unwarrantable assumption of a universal animism or animatism, and invest inanimate things with souls kindred to our own. Evolution, with the genetic relationships and fundamental kinship it implies, accounts for all this intimate emotional appeal. The idea of the universe as a spiritual Society of Souls is a poetical idealised picture, and not in accord with the sober realistic, scientific view of the facts. This is a universe of whole-making, not merely of soul-making, which is only its climax phase. The universe is not a pure transparency of Reason or Spirit. It contains unreason and contradiction, it contains error and evil, sin and suffering. There are grades and gaps, there are clashes and disharmonies between the grades. It is not the embodiment of some simple homogeneous human Ideal. It is profoundly complex and replete with unsearchable diversities and variety. It is the expression of a creative process which is for ever revealing new riches and supplying new unpredictable surprises. But the creative process is not, on that account, issuing in chaos and hopeless irreconcilable conflict. It is for ever mitigating the conflict through a higher system of controls. It is for ever evolving new and higher wholes as the organs of a greater harmony. Through the steadily rising series of wholes it is producing ever more highly organised centres whose inner freedom and creative metabolism transform the fetters of fate and the contingencies of circumstance into the freedom and harmony of a more profoundly co-operative universe. But though the crest of the spiritual wave is no doubt steadily rising, the ocean which supports it contains much more besides the Spirit. Enough for us to know that the lower is not in hopeless enmity to the higher, but its basis and support, a feeder to it, a source whence it mysteriously draws its creative strength for further effort, and hence the necessary pre-condition for all further advance. Thus beneath all logical or ethical disharmonies there exists the deeper creative, genetic harmony between the lower and the higher grades in the Holistic series.

Reference must be made to one more question or set of questions before we conclude. I have said before that the scope of this work is limited, and that it is not intended to deal exhaustively with the entire subject of Holism. But within the limits of the introductory task which I have set myself here, one problem remains to be mentioned. It is the problem of The Whole, the great whole itself as distinguished from the lesser wholes which we have found as the texture of Evolution. In other words, is there a Whole, a Supreme Whole, of which all lesser wholes are but parts or organs? And if there is such a Whole of wholes, how is it to be conceived? Is it to be conceived on the analogy of an organism, as Nature? Or is it to be conceived on the analogy of Mind and Personality as a Supreme divine Personality? Or are both these conceptions inadmissible, and is there some other way of conceiving the system of wholes in their actual or possible synthesis? These are very difficult and thorny questions, but it is clear that we cannot leave the consideration of wholes at the present stage of our argument. For that argument implies clearly something more to complete it, even in the preliminary way which is all that is intended in this work.

Two points arise from the preceding discussion which naturally carry us forward to the consideration of these larger questions. In the first place, where do we fix the limits of a lesser whole? In a whole we have included its field; but how far does this field extend? What limits are there to the field of an inorganic body, or an organism, or a Personality? Leibniz represented each monad as containing or mirroring the whole universe in its own way and from its own particular angle; lower monads, of course, more imperfectly than higher monads; but each in its own degree is a sort of microcosm or miniature universe. In other words, each tiniest least monad is in a sense cosmic and universal. This description would not apply to a field. As we have seen, a field is of the same character as the inner area of the whole, only more attenuated in its force and influence, and the farther it recedes from that area the greater the attenuation; so that the field, though theoretically indefinite in extent, is in effect quite limited in practical operation. When we come to consider a group of wholes we see that, while the wholes may be mutually exclusive, their fields overlap and penetrate and reinforce each other, and thus create an entirely new situation. Thus we speak of the atmosphere of idea, the spirit of a class, or the soul of a people. The social individuals as such remain unaltered, but the social environment or field undergoes a complete change. There is a multiplication of force in the society or group owing to this mutual penetration of the conjoint fields, which creates the appearance and much of the reality of a new organism. Hence we speak of social or group or national organisms. But as a matter of fact there is no new organism; the society or group is organic without being an organism; holistic without being a whole. The mentality of a crowd as distinct from the number of individuals composing it is a good illustration of the changed and reinforced mental field which results from the meeting of many individuals and the fusion and heightening of their conjoint fields. And the more psychic they are, the more they are under the influence of strong passions or carried away by some contagious idea, the more overpowering the common field becomes. The force of the group field is generally out of all proportion to the strength of the idea or the passions in the individual units composing the group. The group field is so to say the multiplication of all the individual fields. The subject falls under the study of social psychology and is referred to here only for the purpose of illustration. We have in such cases an organic situation but not an organism. Groups, families, churches, societies, nations are organic but not organisms.

Taking all the wholes in the world and viewing them together in Nature, we see a similar interpenetration and enrichment of the common field. When we speak of Nature we do not mean a collection of unconnected items, we mean wholes with their interlocking fields; we mean a creative situation which is far more than the mere gathering of individuals and their separate fields. This union of fields is creative of a new and indefinable spirit or atmosphere; the external mechanical situation is transformed into an inward synthetic, " organic " situation or atmosphere. This " organic " Nature seems in certain situations to be alive to us, to stir strange unsuspected depths in us, to make an appeal to our emotional nature which often " lies too deep for words." Thus we come to consider Nature as an organism; we personify her, we even deify and worship her. But the sober fact is that there is no new whole or organism of Nature; there is only Nature become organic through the intensification of her total field. In other words, Nature is holistic without being a real whole.

Nor is it merely we humans, with our intense psychic sensitivity, who feel this appeal of organic or holistic Nature. All organic creatures feel it too. The new science of Ecology is simply a recognition of the fact that all organisms feel the force and moulding effect of their environment as a whole. There is much more in Ecology than merely the striking down of the unfit by way of Natural Selection.

There is a much more subtle and far-reaching influence within the special or local fields of Nature than is commonly recognised or suspected. Sensitivity to appropriate fields is not confined to humans, but is shared by animals and plants throughout organic Nature.

There is a second point which emerges from the foregoing chapters and leads up to the issue now under discussion.

In Chapter VII we have spoken of a general common trend of Evolution, of Evolution as not tacking and veering about, but as moving in one general direction, as keeping a general course and direction through all the endless ages of her voyaging. How is this to be explained? Here again the expedient of personification is often resorted to for the purpose of finding an explanation. It is said that Evolution discloses a grand inner Purpose, that Nature or the universe is purposive or teleological, and that no other category will do justice to the great fact of Evolution as we see it. But if there is purpose there must be a Mind behind that purpose. And thus Mind comes to be personified in Nature as the source of the great evolutionary purpose which the world discloses. Cosmic Teleology spells a corresponding transcendent Personality. Do the facts warrant or necessitate such tremendous assumptions? Would it not rather seem that the whole basis of this reasoning is unsound and false? In all the previous cases of wholes we have nowhere been able to argue from the parts to the whole. Compared to its parts, the whole constituted by them is something quite different, something creatively new, as we have seen. Creative Evolution synthesises from the parts a new entity not only different from them but quite transcending them. That is the essence of a whole. It is always transcendent to its parts, and its character cannot be inferred from the characters of its parts. Now the above reasoning, by which a supra-mundane Mind or Personality is reached, ignores this fact. Such a " Personality " would be creatively new and unlike the wholes which we know and which would constitute its parts. It would be as different at least from human Personality as this again is from mere organism. To call such a new Transcendent Whole by the same name as human Personality is to abuse language and violate thought alike. There is universal agreement with the well-known argument of Kant, that from the facts of Nature no inference of God is justified. The belief in the Divine Being rests, and necessarily must rest, on quite different grounds. From the facts of Evolution no inference to a transcendent Mind is justified, as that would make the whole still of the same character and order as its parts; which would be absurd, as Euclid says. From the facts neither an organism nor a Mind of Nature can strictly be inferred; still less a Personality constituted by both.

Nor is it necessary to make these far-reaching assumptions. There is indeed a great trend in Evolution, but it would be wrong and a misnomer to call that trend a purpose, and worse to invent a Mind to which to refer that purpose. There is something organic and holistic in Nature which shapes her ends and directs her courses. Without forming an organism or a mind the totality of wholes which compose Nature develop an organic field which is sufficient to control her creative movement. As a physical field has its lines of force, so the organic field of Nature, which results from the creative interpretation of all fields of wholes composing her, has its own structural curves of progress. In human society we see how the social field or atmosphere becomes a system of control, a moulding influence to which all incoming members are subject. The individual in society is born into a vast network of controls, and from birth to death he never escapes its subtle toils. The holistic organic field of Nature exercises a similar subtle moulding, controlling influence in respect of the general trend of organic advance. That trend is not random or accidental or free to move in all directions; it is controlled, it has the general character of uniform direction under the influence of the organic or holistic field of Nature.

And there is more. Behind the evolutionary movement and the holistic field of Nature is the inner shaping, directive activity of Holism itself, working through the wholes and in the variations which creatively arise from them. We have seen in Chapter VIII that these variations are not accidental or haphazard, but the controlled, regulated expression of the inner holistic development of organisms as wholes. There is Selection, and thus direction and control, right through the entire forward movement, not only in the origin of variations but also at the various subsequent stages of their "selection," internal and external. This organic holistic control of direction, this inner trend of the evolutionary process is really all that is meant by the metaphor of Purpose or Teleology as applied to Nature or Evolution. To infer more is in effect to make the mistake of spiritual Idealism and to apply later human categories to the earlier phases of the evolutionary process.

Thus it is that when we speak of Nature or the Universe as a Whole or The Whole, we merely mean Nature or the Universe considered as organic, or in its organic or holistic aspects. We do not mean that either is a real whole in the sense defined in this work. We have seen that the creative intensified Field of Nature, consisting of all physical organic and personal wholes in their close interactions and mutual influences, is itself of an organic or holistic character. That Field is the source of the grand Ecology of the universe. It is the environment, the Society—vital, friendly, educative, creative—of all wholes and all souls. It is not a mere figure of speech or figment of the imagination, but a reality with profound influences of its own on all wholes and their destiny. It is the οἶκος, the Home of all the family of the universe, with something profoundly intimate and friendly in its atmosphere. In this Home of Wholes and Souls the creative tasks of Holism are carried forward. Without idealising it unduly we yet feel that it is very near and dear to us, and in spite of all antagonisms and troubles we come in the end to feel that this is a friendly universe. Its deepest tendencies are helpful to what is best in us, and our highest aspirations are but its inspiration. Thus behind our striving towards betterment are in the last resort the entire weight and momentum and the inmost nature and trend of the universe.

I have now reached the end of my argument. The reflections embodied in this work lie far removed from the busy and exciting scenes in which most of my life has been spent; and yet both of them tend toward the same general conclusions. It has been my lot to have passed many of the years of my life amid the conflicts of men, in their wars and their Council Chambers. Everywhere I have seen men search and struggle for the Good with grim determination and earnestness, and with a sincerity of purpose which added to the poignancy of the fratricidal strife. But we are still far, very far, from the goal to which Holism points. The Great War—with its infinite loss and suffering, its toll of untold lives, the shattering of great States and almost of civilisation, the fearful waste of goodwill and sincere human ideals which followed the close of that vast tragedy—has been proof enough for our day and generation that we are yet far off the attainment of the ideal of a really Holistic universe. But everywhere too I have seen that it was at bottom a struggle for the Good, a wild striving towards human betterment; that blindly, and through blinding mists of passions and illusions, men are yet sincerely, earnestly groping towards the light, towards the ideal of a better, more secure life for themselves and for their fellows. Thus the League of Nations, the chief constructive outcome of the Great War, is but the expression of the deeply-felt aspiration towards a more stable holistic human society. And the faith has been strengthened in me that what has here been called Holism is at work even in the conflicts and confusions of men; that in spite of all appearances to the contrary, eventual victory is serenely and securely waiting, and that the immeasurable sacrifices have not been in vain. The groaning and travailing of the universe is never aimless or resultless. Its profound labours mean new creation, the slow, painful birth of wholes, of new and higher wholes, and the slow but steady realisation of the Good which all the wholes of the universe in their various grades dimly yearn and strive for. It is the nature of the universe to strive for and slowly, but in ever-increasing measure, to attain wholeness, fullness, blessedness. The real defeat for men as for other grades of the universe would be to ease the pain by a cessation of effort, to cease from striving towards the Good. The holistic nisus which rises like a living fountain from the very depths of the universe is the guarantee that failure does not await us, that the ideals of Wellbeing, of Truth, Beauty and Goodness are firmly grounded in the nature of things, and will not eventually be endangered or lost. Wholeness, healing, holiness—all expressions and ideas springing from the same root in language as in experience—lie on the rugged upward path of the universe, and are secure of attainment—in part here and now, and eventually more fully and truly. The rise and self-perfection of wholes in the Whole is the slow but unerring process and goal of this Holistic universe.

INDEX

ABSOLUTE Values as wholes, 107
Absolutists and "the whole," 100, 102, 110; their sterile monism, 108
Abstraction, the error of, 15, 20, 27
Acceleration and gravitation, equivalent expressions, 28-9, 30
Acquired characters, inheritance of, 194, 201
Acquired experience, inheritance of, 205 n.
Action, the physical basis of the universe, 31, 41, 44, 51, 56, 326, 327, 328; its structural character, 41-2, 44, 327
Adsorption, 46
Alchemists, their guesses not far from the truth, 42-3
Alga and fern, possible bridge between, 74-5
Algonkian mountains, age of, 42
Alternation of generations, basis of, 73-4
Analysis, the error of, 19-20, 27
Anaxagoras, 229
Aniline dyes, 57
Animals: the psychical development manifested in Sexual Selection, 13-14; plants and, their similarities and divergences, 70, 71, 73, 75; co-ordination and self-regulation in, 77, 82, 105, 106, 230, 231; power of self-healing in, 80-81; regarded as wholes, 82, 83, 105, 106; animals and humans, heredity and educability, 252, 274
Armstrong, Dr. E. F., 47
Artistic creations as wholes, 98, 105-6
Associative memory of white mice, experiments on, 205 n.
Atom, the, 37, 40, 61, 71, 83-4, 97, 173; importance of placing and spacing of atoms, 37-8; not static but active in the Space-Time continuum, 38; the conquests of the New Physics, 38-9; theory of Rutherford and Bohr, 39-41; nucleus, electrons, and quanta of radiation, 39-40, 41, 49; its planetary structure, 39, 40, 45, 48, 83-4; the proton, 40, 41; as a potential source of energy, 42; the artificial breaking up of matter and the transmutation of metals, 42-3; external properties dependent on internal structure, 43-4; the basis of the hypothesis of its structure, 48-50; a creation of Holism, 128, 153, 228, 229, 319-20; discontinuity in its structure, 331
Atomic numbers, 40-41
Attention, development of power of, 236, 237, 240, 241

Baly, Professor, 69
Beauty, its holistic basis, 221, 222
Becquerel, 23, 38
Bergson: his philosophy of Evolution summarised and examined, 92-5; the principle of Duration, 92-3, 94, 95, 99, 115; the Intellect, 93-4, 95, 111
Berkeley, 270, 272
Bio-chemical mechanisms, 151
Bio-chemical wholes, 151-2, 154, 168, 169. See Organisms.
Bio-chemistry, recent advances in, 72
Biographies as aids to the science of Personology, 284-9, 294
Biological wholes, 100, 101, 104
Biology: and a new concept of life, 3, 5; Mechanism and, 4, 5, 11, 109, 113-14, 121; latest advances in, 5, 7; new syntheses more important than specialisation, 5, 7, 11; Sexual Selection in Organic Evolution, 14; its study of Evolution and the cell, 61; value of the concept of the whole to, 110, 221, 259; value of the concept of

"fields" to, 113-16; Vitalism and, 160; the germ-cell theory of Variation, 190, 197-203; Mendelism, 194, 195-6; Personality compared to biological sports, 276

Body, the: needless confusion over interaction of mind and, 157, 158; early Christian controversy on immortality, 158; Descartes and the relation of mind and, 161, 163, 170; the laws of thermodynamics and the principles of mind and, 163-70; Holism on the action of "life" on, 172-81; the Subject-Object relation of mind and, 239; in relation to Personality, 264-7, 279, 280; relations of spirit and, 264, 266-7; conventionally degraded by morbid religious spirit, 265; rehabilitated by modern science, 266; relation of mind and, in Personality, 267-71

Body-cells differentiated from germ-cells, 198-9, 201; possible reciprocal influence of body-cells and germ-cells, 203, 204, 205, 206, 219

Bohr, Professor Niels, 39

Bower, Professor F. O., 207 n

Brain, the: its holistic functions, 143; mind and, 244, 253

Breathing, physiology of, 140

Brown, Robert, 62

Cage, closed, Einstein's illustration, 28-9

Carbon dioxide, its transformation in the plant, 68, 77

Carlyle, arrested development of his inner self, 286

Catalysis, use of colloids in, 46, 68

Causation, rigid concept of, in 19th-century science, 9, 16-17, 155, 156; and the idea of "fields," 17-19; Holism and, 126-7, 135-6, 137, 138, 305, 306-7, 308; freedom creative within the process of, 137-9, 306-7

Cell, the, 61, 63, 64, 83, 84, 97, 153, 221, 228, 229, 230; colloidal system of plant cell, 46-7, 65; enzyme action in, 46-7, 68-9; the chromosomes, 53, 62, 70-71, 73, 202, 220; history of study of cells, 62-3; structure and functions of cells, 62, 63-9; the cell wall, 62, 65; cell-divisions, 62, 63, 70-71, 72, 73-4, 75; attempt to explain physical mechanism of heredity from, 63; some central control of its functions implied, 66, 67-8, 77, 78-83, 84, 88, 191; its origin possibly electrical, 70-75; reproduction by reduction division, 73-4, 75; single-content and double-contents cells, 73-4; differentiation in cells, and divergence of plant and animal forms, 75-6, 80; the co-operation and co-ordination of cells towards a whole, 77-84, 96-7, 191, 214-15, 230, 320; the power of restitution of damaged cells, 80-81; body-cells differentiated from germ cells, 198-9, 201; possible reciprocal influence of body-cells and germ-cells, 203, 204, 205, 206, 219; germ-cell theory of Variation, 188, 190, 191, 194, 197-205, 220.

Cell-division, process of, 63, 70-71, 72-4, 75; its electrical character, 70, 71, 72

Characterology, 282

Chemical affinity, accounted for, 38, 43-4; the selectiveness of matter in its colloid state possibly related to, 57; Holism and, 319

Chemical compounds, holistic character of, 104, 105, 106, 122, 128, 133, 173; Mechanism and chemical combination, 150

Chemistry, 54; the analysis of the constitution of matter, 37; the importance of "structure," 37, 38; two types of chemical change, 47-8

Chlorophyll, its part in plant life, 47, 58, 68, 71, 76-7

Christianity and the evolution of the idea of Personality, 283

Chromosomes, 62; differences in, and organic variations, 53; their behaviour in cell-division, 70-71, 73; hereditary characters carried by, 200, 220

Clerk-Maxwell, 166-8, 169, 170, 172

Colloid state of matter, 45-7, 57-8, 68, 162; distinctive of all life-forms, 46; colloidal system of plant cell, 46-7, 65; distinctive properties of colloids, 46-7, 68-9; enzyme action, 46-7, 68-9; anticipates processes and activities of life, 57-8

INDEX

Compounds, unstable equilibrium and formation of, 44
Conation, development of capacity of, 237, 240, 242
Concepts and their "fields," 17-19; holistic unity of conceptual system, 244, 246, 247
Consciousness: the closed system of physical science denied by, 155-7, 165, 169; the development of, 236, 237-8, 239, 240, 241, 250, 320; the Subject and the Object in, 238-40, 292, 307; and the "field" of mind, 253; its freedom and spontaneity, 306, 307, 308, 309, 320
Conservation of energy, law of, and the systems of life and mind, 163, 164, 165-70, 305
Co-operation and co-ordination, holistic, 77-84, 105, 214-15, 220, 230-32; in the cells, 77-84, 96-7, 191, 230, 320; in variations, 208-9, 210, 220; mental processes crude in comparison, 231
Creation: two views of the creation of the universe, 88-90, 129; an unintelligible sense of the word creation, 128, 131; holistic creation, 137. *See* Creative Evolution.
Creative Evolution, 89-96, 128, 137, 141, 171, 217, 275, 328; modern belief in, and its implications, 9-11; science and philosophy brought together by, 90-92; as structure plus principle, 91, 92; Bergson's system summarised and examined, 92-5; its holistic character, 98, 99, 102, 129-31, 137, 141-2, 143, 211, 212-15, 217, 321 *n.*, 323, 325, 328, 332, 335, 342-3; fundamental issues raised by, 132-40; the lower unit always the basis of the next higher, 153-4, 169, 175, 178, 331; beyond the scope of Mendelism, 196; the germ-cell theory of Variation, 188, 190, 191, 194, 197-205; negative aspect of, 215-16; Naturalism irreconcilable with, 329-30; Spiritualism irreconcilable with, 330-31, 332, 333, 334, 335; discontinuity of its progress, 331-2; Monadology and, 332-4; Pluralism and, 334-6; and the idea of a Supreme Whole, 341-3

Creative Evolution (Bergson), 93, 94, 96
Creativeness: of mind, 32, 90, 246, 250-52, 336; of matter, 55-8, 89-90, 96, 128; of thought, 90; of Holism, 101, 107-8, 128-36, 142-4, 178, 181, 221, 222, 273, 304, 305; fundamental issues raised by concept of, 132-40; reaches its maximum in Personality, 275; of Personality, 304-5
Crowd mind, the, 339-40
Crystal structure, 38, 45; the lattice pattern, 45; the unit body, 45; process of its growth, 67
Curves, the pathways of all events in the Space-Time universe, 30, 31
Cytology, present study of, 63. See Cell.
Cytoplasm, 65

Darwin, 6, 11, 23, 185-6, 187
Darwinism: a new view-point in respect of existing knowledge, 6; the theory of Descent and Natural and Sexual Selection not mechanical but psychical, 11-16; Holism and, 182-223; the Darwinian theory summarised, 186-8, 192-4, 195; Variation, 186, 187, 188, 190, 192-4, 218; Natural Selection, 186, 187-90, 192, 193, 218, 219; the Neo-Darwinians, 190-2, 194-203, 218, 220; co-ordination and co-adaptation of organs and characters not explained by orthodox Darwinism, 208-9; Holistic Selection and Natural Selection, 209-10, 211-19
De Vries, 53, 63, 189, 194, 196-7
Descartes, on the relations of mind and body, 161, 163, 170
Descent: Darwin's theory of, 11-16, 186, 187 (*See* Darwinism); the operation of Radioactivity compared to that of Organic Descent, 52-3, 54; creative Evolution and, 129, 131
Determinism, holistic, 308, 310, 311
Development and Purpose (Hobhouse), 247
Dissipation of energy, law of, and the relations of life and mind, 163, 164, 165-70
Doncaster: *Introduction to Study of Cytology*, 320 *n.*

INDEX

Double-contents cell, generation of, 74
Driesch, Professor Hans, 171-2
Ductless glands, holistic organs, 143, 205
Duration, Bergson's principle of, 92-3, 94, 95, 99, 115
Durkhen, experiments of, 207 n.
Duty, Personality and, 314

Ecological modifications leading to variations, 207, 208, 210, 218, 219, 238
Ecology, science of, 218, 340-41, 343. See Environment.
Einstein, 6, 7, 239, 240; the new view-point of Relativity, 6; publishes *General Theory of Relativity*, 23; the theory capable of being put simply and intelligibly, 24-5; its mathematical origin, 24, 25-7; the Ten Equations, 25, 185; the old mechanics and the new, 25-32; motion never absolute, 25-6: the *Special Theory of Relativity*, 26-7; his application of the concept of the Space-Time continuum, 28-32, 33-4; the illustration of the closed cage, 28-9; the idea of the inertia of matter destroyed, 28; the subjective and objective in the Space-Time synthesis, 33, 34
Élan vital, 99
Electro-magnetism, an instance of the selectiveness of matter, 161-2
Electrons, 39, 40, 49; and the nucleus, 39; combinations of protons and, 40; external properties of the atom decided by number and grouping of, 43
Elements, the, 41, 52; their nuclei, 40; atomic numbers, 40-41; the Periodic Table, 41, 42, 52; their spontaneous breaking up in Radio-activity, 42, 52; artificial destruction and transmutation of, 42-3, 52, 53; external properties the expression of internal structure, 43-4
Elimination of the unfit, 11-12, 13. See Natural Selection.
Embryo: formation of, by cell-division, 62; phylogeny repeated in ontogeny, 74, 115
Emergent Evolution (Morgan), 321 n.
Ends, the realm of, 259

Energy: the "field" of an object, 17-18, 112, 327-8; matter simply a form of, 31, 41, 44, 51, 56, 326, 327, 328; intimate relation between structure and, 41-2, 44, 51, 55, 56, 327; potentially available by artificial breaking up of matter, 42; first functional in the cell, 64, 65; holistic use and control of, 105, 106, 107; laws of, and principles of life and mind, 155, 163-70, 305
Entelechy, theory of, 171-2, 177, 269
Environment: a confused complex concept, 113; the organism largely independent of, 136-7, 138, 300, 340; Variation and, 207, 208, 210, 218, 219; consciousness increases influence of, 238; mind as creator of, 251; social inheritance borne by, 251-2; Personality and, 300, 301-4
Enzymes, action of, 46-7, 68-9
Epistemology, Personality and, 277
Equilibrium: of the atom, and its external properties, 43-5; instability and readjustment of fundamental structures of Nature, 173-4; the same rhythm in the structure of life, 174, 176, 177-8, 216, 235; persistent overbalance caused by Holism, 178-9, 214; tension and selective compensation as the source of mind, 235-6
Ether, hypothesis of, 17, 269, 323-4
Ethics: individualism and ethical problems, 241, 246; ethical character of Personality, 294, 295, 298, 299-300, 304-5, 311, 312, 313, 314-16
Euclidean geometry and the theory of Relativity, 25, 29, 30; suitable to the Newtonian and Kantian conception of Space and Time, 32
Evolution, 2-3, 8-10, 88-90, 141, 171, 217, 323, 325, 328, 331-2; Darwinism, 6, 11-16, 186-90, 191-4, 218; the 19th-century battle over, 8-9; the modern belief in Creative Evolution, and its implications, 9-11, 23, 89-90, 129-35, 137 (See Creative Evolution); mechanistic conceptions strengthened by Darwinism, 11-16, 189, 190; its tendency to hark back to simpler types, 54; the idea applicable to matter,

56-7, 89-90; the position of the cell in, 64-5, 83, 84; cell differentiation and the divergence of plant and animal forms, 75-7; creative Holism the motive force behind Evolution, 98-99, 102, 105, 107-8, 129-35, 137-9, 141-3, 178, 216, 220-3, 230, 273, 320, 321 n., 323, 325, 326, 328; the lower unit the basis of the next higher unit, 153-4, 169, 175, 178; life and mind the two great *saltus* in, 154, 271; its persistent trend not accidental, 178, 341, 342; Neo-Darwinian theories, 188, 190-92, 194-203; the Germ-cell theory, 188, 190, 194, 197-205; Mendelism, 194, 195-6; the Mutation theory, 194, 196-7; experimental Evolution and its limitations, 194, 195-6, 200, 217; Holistic Selection in, 211, 212-15, 220; its negative aspect, 215-16, 221-2; interaction of internal and external factors in, 218-20; the place of mind in, 226, 228, 229-35, 248-9, 251; the psychic and the organic in, 230-32, 234, 235-8, 240-41, 243-4, 249-53, 273-4; its advance towards individuality and Personality, 232-3, 275, 276, 282, 320; Naturalism and, 329-30; Monadology and, 332-4; Pluralism and, 334-5; purposive view of, justifies no inference of a Supreme Mind, 342

Evolution in the Light of Modern Knowledge (Bower), 207 n.

Experience: interaction of matter, life, and mind confirmed by, 2, 3, 155-6, 157; its plasticity and the rigidity of our concepts, 17, 21, 23; the Space-Time continuum of events in, 27-8, 33, 34; the Subject-Object relation in, 32-3, 94, 238, 239, 277; Bergson on, 92-3; the duality of the mind and, 238, 239, 240, 245, 246-51; scientific interpretation of, 247-8, 250; influence of past on present, 254; Personality as the Subject of, 277, 292

Females, emotional sensitiveness of, implied by the principle of Sexual Selection, 13, 14, 222-3

Ferments, action of, in protoplasm, 47, 68

Ferns, gametophytes of, 74-5; modifications and variations in, 207 n.

Fibro-vascular system of plants, 76, 78

Fields, value of the concept of, 17, 18, 110, 111-14, 328, 339; applied to concepts, 17-19; things and their fields, 17, 112, 113-14, 327-8; the Space-Time continuum the field of the material universe, 31-2, 34, 114, 327; the field of matter dependent on its internal structure, 44, 110, 112-13; colloidal surfaces as fields, 47; wholes and their fields, 110, 112, 113-16, 335, 339, 340; organisms and their fields, 113-16; the field of the germ-cell, 191, 204, 206-8, 219; the field of mind, 228, 253-9; the field of Nature, 340-41, 342, 343; group fields, 339-40

Force, doubtful validity of concept of, 160; laws of, and the principles of mind and body, 163-70

Freedom: of the organic whole, 137-9, 306, 307, 309; of the Personality, 139, 274, 275, 276, 304, 305, 309-12, 314, 315; of the mind, 250, 258-9; the rule of the universe, 307-8; not limited to the will, 307, 309, 310

Function and structure, their relation in wholes, 104, 106, 107, 112, 122-4

Future, the, an operative factor in the activity of the mind, 258-9

Galileo, 25
Gametes, the, 74
Gametophyte generation, the, 74
Gases, result of internal equilibrium in the atoms and molecules, 44, 45
General Theory of Relativity (Einstein), 23, 24
Generalisation, the error of, 15, 20
Genes, 53, 200; struggle for existence assumed among, 201-2
Genetics, 563; theories of Variation, 53, 194-205, 220
Geological age measured by Radio-activity, 42
Germ-cell theory of Variation, 188, 190, 191, 194, 197-205, 220; the field of the germ-cell, 191, 204,

206-8, 219; possible reciprocal influence of body-cells and germ-cells, 203, 204-5, 206, 219
Glutathione, 72
God: as the medium of the interaction of body and mind, 270, 272
Gold, transmutation of Mercury into, 42-3
Gravitation: Relativity and, 25, 28-31, 185; Einstein's closed cage illustration, 28-9; acceleration and, equivalent expressions, 28-9, 30; as the curved structure of the real Space-Time world, 30-31; Newton's Law of, 184-5
Grew: study of plant cells, 62
Guelincx, 272

Habit and hereditary modifications leading to variations, 203, 204, 206-8
Hæmoglobin, 58
Haldane, Professor, 140, 177
Hegel, 88
Helium atom, the, 40, 42; Helium nuclei, 40, 42; emission of Helium used as a geological clock, 42; transmutation of elements by expulsion of Helium atoms, 42, 53
Heredity, 63; the organism and its field, 114-16; Variation infinitesimal compared to, 141; inheritance of acquired characters, 194, 201; inheritance of modifications, 194, 201, 202-3; inheritance of mutations, 194, 196-7, 206; Mendelism, 195-6; the germ-cell theory and, 198, 199, 200; inheritance of acquired experience, 205 n.; educability and, in the human, 252, 274; the hereditary past in Personality, 255, 273, 274; Holism and, 255, 273, 274, 332
Hobhouse, Professor L. T., 247
Holism, 98, 107, 116-17, 148, 180; co-operation and co-ordination in cell activities, 77-84, 96-7, 191, 214-15, 230, 320; the fundamental whole-making tendency of the universe, 82, 84, 97-8, 99, 100, 107, 108, 143, 179, 181, 307-8, 319-21, 326, 329, 335, 337-8, 344-5; its progressive phases, 96-7, 105, 106-7, 180, 335; ideal wholes, 98, 105-6, 107, 144, 222, 241, 243, 259, 294, 305, 311, 312, 313, 314-16, 329, 344-5; the motive power of creative Evolution, 98, 99, 102, 105, 107-8, 129-31, 137, 142, 143-4, 215, 321 n., 323, 325; the source of all Values, 107, 144, 221, 241, 243, 259, 305, 335, 344-5; creativeness of, 107-8, 128-36, 142-4, 178, 181, 221, 222, 273, 304, 305; value of the concept of, 108-10, 148, 259, 260; bridges the gaps between matter, life, and mind, 108-9, 121, 158, 174-6, 178, 179, 320-21; Science and, 109-10, 142, 148, 221, 248, 321-5; suggested substitution of notion of Holism for that of life, 110, 160-61; as a concept, a factor, and a theory, 116-17, 158; functions and categories of, 120-44; and the idea of causality, 126-7, 135-6, 137, 138, 305, 306-7, 308; Freedom and, 137-9, 274, 304, 305, 306, 308-14; inner co-ordination and self-regulation by, 140-42, 143, 214-15, 230-31, 232, 233, 234, 240, 242, 319-20; structural character of, 143-4, 153, 175, 180-81; Mechanism and, 147-81, 292, 329; Personality the supreme expression of, 151-2, 153, 263-4, 267, 273, 283, 284, 292, 293, 295, 304, 308, 316, 320, 329; and life and its action on the body, 172-9; overbalance of equilibrium in all structures towards, 178-9, 214; Darwinism and, 184-223; Variation as explained by, 191, 206, 209-15, 332, 342-3; Holistic Selection, 211, 212-15, 220; repressive aspect of, 215-16; spiritual aims of, 216, 222, 337-8, 344-5; mind as an organ and expression of, 222, 226-60, 264, 292, 320, 325; individuality and universality, two tendencies of, 232, 233-5, 238, 240-41, 242, 243, 245-6, 319; development of attention and consciousness, 237; the Subject-Object relation in, 238, 239, 240, 293, 295; holistic aspects of subconscious mind, 254-5, 257, 279-80; the senses and, 255-6, 284; purpose and, 258-9; body and spirit reunited in, 266-7; and body-and-mind relation in Personality, 267-73, 277, 279; itself the real actor in Personality, 271-6, 283,

INDEX

299; self-realisation the aim of, 294, 314-16; will and, 295; the ideal of Purity and, 303-4, 312, 314-15; moral discipline in scheme of, 313, 314; the holistic universe, 319-45; unverifiable, 321, 322, 324-5; Spiritualism and, 332, 333, 334-6; Monadology and, 332-4, 335; Pluralism and, 334, 335-6; the Supreme Whole in, 338, 341-2, 343.
Holistic Selection, 211, 212-15, 220
Hooke, Robert, 62
Hormones, functions of, 205
Hybridisation, 217
Hydrogen atom, the, 40, 41-2

Ideal Man, the, 312-13
Idealism, a denial of creative Evolution, 330-1, 332, 334, 335, 337, 343
Ideals, Holistic, 98, 105-6, 107, 144, 222, 241, 243, 259, 294, 305, 311, 312, 313, 314-16, 329, 344-5
Immortality, early Christian controversy on, 158
Individual, the, and the race, differentiation of development of, 198-9, 201-3, 204
Individuality: fundamental in Nature, 83; mind and the holistic process of individuation, 139, 140, 232-4, 238-40; consummated in the human Personality, 233, 241, 245-6, 274-5, 276, 277, 278-9, 284-5, 289; in its higher developments, 240-46; the individual and his social environment, 245, 251, 252, 339, 342; the individualist mind and the universal mind, 245-6; memory as the basic bond of individuality, 254; the study of the individual Personality in biography, 284-9
Inert elements, and internal equilibrium of the atom, 43, 44
Inorganic, the: vanishing fixity of inorganic elements, 23-4; its conversion into organic at colloidal surfaces, 47, 58; chemical combination and structure in inorganic chemistry, 47-8; the cell the real distinction between organic and, 64
Instinct: in Bergson's philosophy, 94; and the development of mind, 237

Intellect, the: in Bergson's philosophy, 93-4, 95, 111; effect of its selectiveness, 111, 112; development of, 294-5
"Intro-action" of body and mind, 270
Intuition, in Bergson's philosophy, 94
Inverse Square, Newton's law of Gravitation, 185
Iodine, its effect on the thyroid gland, 279
Isomerism, 38

Judgment, synthetic, 259-60

Kammerer, 207 n.
Kant: his conception of Space and Time, 32, 33, 94; on the creative action of the mind, 32-3; and the Newtonian system, 185; his realm of Ends, 259; his "synthetic unity of apperception," 259-60, 271; man a legislative being, 296; the concept of Necessity, 308; no inference of God justified from the facts of Nature, 342
Keats, 336

Lamarck, reaction towards, 188
Language, a social instrument, 245, 251
League of Nations, the, 344
Leibniz: pre-established harmony of, 34, 89, 269-70, 272, 333, 334; his reply to Descartes, 164, 167; Holism and the Monadology of, 332-4, 335, 339
Life: apparent separateness of life, mind, and matter not founded in fact, 2, 3, 21, 50; new concept of, needed, 3, 4, 5; mechanistic view of, 4, 12, 109, 110, 155, 179-80; thought and, 4, 157; vagueness of present concept of, 5, 16, 109, 120, 158-9; its development from matter, 7-8, 10-11, 174-6; life and mind true operative factors in Evolution, 15-16, 324-5, 332; chemical structure of its mechanism, 48; in the cell, on its way to mind, 66, 77; electrical energy of sun and the origin of, 71, 72; character of wholeness in, 77-84, 97; overflow of life, mind, and matter into each other's domain,

85, 97, 108, 227, 228, 270; the gaps between life, mind, and matter bridged by Holism, 108-9, 121, 158, 174-6, 178, 179, 320-21; the concept of the whole preferable to that of, 109-10, 160-61; degrees of Freedom in, 139; the laws of thermodynamics and the principles of life and mind, 155, 163-70; development of mind from, 151-2, 154, 178, 180, 231-4, 235-8; life and mind not independent entities, 157-9, 170; the Vitalistic hypothesis criticised, 159-61, 171-2, 177; power of selection and self-direction in, 162-4, 165, 167, 168, 170-71, 176-7, 235, 237, 249; unity and inter-action of life and mind, 170-71; the theory of Entelechy and, 171-2, 177; life and its action on the body, Holism and, 172-9; a new structure based on those of the physico-chemical order, 174-6, 178, 179, 181, 230, 335; the rhythmic equilibrium in, 174, 176-9; Goethe on its purpose, 315; Naturalism and, 329

Light: its velocity, and the principle of Relativity, 26-7; the curvation of, 30; accounted for by quanta of radiation released by electrons, 39-40; light effects as basis of theory of atomic structure, 49-50

Liquids: and internal equilibrium of the molecule, 44-5; molecular structure of, 45, 64

Males, the operation of Sexual Selection limited to, 13
Malpighi, 62
Man: a psycho-physical whole, 152; his mind and his environment, 251-2; the rôle of structure not very prominent in, 252-3; hereditary past in mind of, 255, 274; heredity and educability in, 274; "a legislative being," 296; the Ideal Man, 312-13; Naturalistic view of, 329-30
Materialism, its unjustified inferences, 8, 10, 15; its struggle with Spiritualism over Evolution, 8-9, 11
Matter: apparent separateness of matter, life and mind, not founded in fact, 2, 3, 21, 50; new concept of, needed, 3, 4-5, 10, 11, 16, 50-52, 56, 57, 61, 96; thought and, 4; development of life and mind from, 7-8, 10-11, 51, 52, 53-4, 55, 57, 58, 174-6; 19th-century materialism, 8, 9; Evolution and the new concept of, 10-11, 23; its field the structure of the Space-Time universe, 31, 44, 112, 327; a form of Energy or Action, 31, 41, 44, 51, 55, 56, 326-7; recent advances in the knowledge of its constitution, 37-50, 95, 162; structural character of, 37, 38, 44-5, 51, 55, 56, 95, 162, 173-4, 230, 231, 331; the proton possibly the fundamental form of, 41; artificial breaking up of, greatest potential source of energy, 42; internal structure and external properties of, 44-5, 112; its behaviour in the colloid state, 45-7, 57-8, 162; selectiveness of, 47, 57, 161-2, 235; creativeness of, 55-8, 89-90, 96-7, 128; in the cell, 64, 65; overflow of matter, mind, and life into each other's domain, 85, 97, 108, 227, 228, 270; Holism and the disappearance of the gulf between life, mind, and matter, 108-9, 121, 158, 174-6, 178, 179; life a complex structure of, 174-6, 178, 179, 230, 335; spirit and, Spinozist position, 321 n., 334-5
Mechanical system, holistic system distinguished from, 103-4, 125-6, 127, 133, 141, 305
Mechanics, 25; the new system of Relativity, 25-34
Mechanism, 149, 152; invades the domain of life, 4, 5, 12, 109, 110; biology and, 4, 5, 11, 100, 109, 113-14, 121; the system shaken by the concept of creative Evolution, 10-11, 132-3; strengthened by a misconception of Darwinism, 11-16, 189, 190; its recent domination of science, 89, 100, 103, 322, 324-5; wholes not mechanical systems, 103-4, 125-6, 127, 133, 141, 305; Holism and, 147-181; the concept of Holism transforms and transcends the mechanistic system, 147, 148-52, 179-81; man and, 152; an early phase of Holism, 153, 180, 329; inadequate for mod-

ern physiology, 177; its view of Variation arbitrary and misleading, 190-91, 214, 215, 220; unable to cope with Personality, 292

Memory, 254; associative, 205 n.; hereditary, 255

Mendel, Abbot, 195

Mendelism, 63; its theory of Variation, 194, 195-6

Mercury, its possible transmutation into Gold, 42-3

Metabolism, the process of, 67, 78, 101, 129, 165, 191, 300; the same power necessary to the Personality, 301-4

Metaphysics, Holism and, 260

Mice, associative memory in, 205 n.

Milton, 288

Mind: apparent separateness of mind, life, and matter not founded in fact, 2, 3, 21, 50; new concept of, needed, 4-5, 10, 11, 16; general acceptance of physical basis of, 7-8, 10-11; life and mind, true operative factors in Evolution, 15-16, 324-5, 332; mind and the Space-Time universe, 32, 33, 254; Kant on, 32-3; creativeness of, 32, 90, 246, 250-52, 336; life in the cell on its way to, 66, 77; its development in animals, 77; the overflow of mind, life, and matter into each other's domain, 85, 97, 108, 227, 228, 270; structure and, 94, 95, 252, 253; gaps between mind, life, and matter bridged by Holism, 108-9, 121, 158, 174-6, 178, 179, 320-21; its development from life, 151-2, 154, 178, 180, 231-4, 235-8, 336; the closed system of physical science and, 155, 156, 161; life and mind not independent entities, 157, 158, 170; the laws of thermodynamics and the principles of life and mind, 155, 163-70; power of self-direction in, 162, 163-4, 165, 167, 168, 171, 250-52, 270; unity and interaction of life and mind, 170-71; the theory of Entelechy and, 171-2, 177; psychology and, 227, 228; the field of, 228, 253-9; Personality and, 229, 233, 235, 264, 267-73, 277, 279; crude as compared with organic co-ordination and self-regulation, 231; lines of advance of its evolution, 231-5; as an organ and expression of Holism, 222, 226-60, 264, 292, 320, 325; individuation and, 233, 234-5, 241, 244-6, 248; organisation and central control of, 233, 234, 248, 250, 292; development of attention and consciousness, 236, 237-8; duality of, the Subject-Object relation, 238-40; mind and body not independent, 239, 267-73, 277, 279; as a rebel against universality, 241, 244, 246, 248; Reason and, 243, 244, 247-8; the Self conquered by, 244-6; Science the proudest achievement of, 247-8, 249, 250; its enrichment of the universe, 248-51; and its environment, 251-2; the subconscious mind, 254-5, 257, 279-80; influence of the past on, 254-5, 257, 336-7; the senses and, 255-7; telepathy and, 257; influence of the future on, 258-9; purpose the highest manifestation of its activity, 258-60; body and mind in Personality, 267-73, 277, 279; Naturalistic view of, 329; the assumption of a Supreme Mind, 341-2

Minkowski, 27

Modifications, Darwin's theory of, 192, 193; inheritance of, negatived by Weismann, 194, 201, 202-3; possibly the conditions precedent to Variation, 203, 204-5, 206, 207, 212, 219

Mohl, von, 62

Molecule, the, 37, 153; importance of the placing and spacing of its atoms, 37; combination of atoms into molecules rests on unstable internal equilibrium, 43, 44; molecular structure of liquids, 45, 64; lattice pattern in crystal structure, 45; in the colloid state, 46

Monadology and Holism, 332-4, 335, 339

Monistic conception of the universe furthered by Holism, 108-9

Moral character and the influence of Holism, 299-300, 313, 314

Morgan, Professor Lloyd, 321 n.

Motion: Newton's laws, 25; the Einstein theory, 25-34; never absolute, 25, 27; stationariness an illusion, 25-6; the co-variation of Space and Time, 26-30; the Space-Time continuum applied to, 28-30

Multicellular organisms: reproduction by cell fusion, 73; reproduction by reduction division, 73-4, 75; formation of the earliest, 75-6

Mutations: De Vries' theory of, 53, 194, 196-7, 206; of matter to life, 57; Darwin's theory of, 192, 193; exceptional, 197, 218; of body to mind, 270-71

Natural Science reunited with psychology, 240

Natural Selection, erroneous mechanical view of, 11-12, 14, 15, 19, 189, 190; fundamentally psychical, 14-15, 222-3, 313; Darwin's theory of, 186, 187-8, 189-190, 192, 193, 206, 219; co-operation between Variation and, 192, 193, 210-16; operative within the germ-cell, 201-2; of small variations, 205-8, 210-14; Holistic Selection and, 206, 209-16; its limitations, 215; co-operative and helpful rather than murderous, 218, 219

Naturalism: and the principles of life and mind, 154-5, 156, 157, 169, 170, 171; Holism and, 329-30, 331; irreconcilable with creative Evolution, 329-30, 332, 334

Nature: errors in the observation of, 19-21; new view of, 23-4, 266; her high-speed internal energies, 51; the concept of Holism and the explanation of, 95, 96, 108, 131-5, 222, 342-3; mind and, 96, 97, 156, 229, 336, 342; wholes as the real units of, 99, 100, 108, 110; value of the Space-Time integration to the understanding of, 111, 122, 173; fields in, 111, 112-13, 340-41; the concept of creativeness and, 131, 132, 133-5; the closed system of physical science, 153-4, 156; life a new structure of her holistic physico-chemical structures, 173, 174-9; warfare not the rule in, 218, 219; the emotional appeal of, 336-7, 340, 343; holistic, not a whole, 340, 342, 343; teleological view of, 341-2; the holistic Field of, 342, 343; the inference of a Supreme Mind behind, 341-2, 343

Necessity: in the closed system of physical science, 155; the limitations of its power over wholes, 305-6; the concept not grounded in reality, 308

Neo-Darwinians and the theory of Variation, 190-92, 194-205

Newton, 7, 25, 32, 164, 185, 186; his First Law and the inertia of matter, 32, 51; his conception of Space and Time, 32, 33; his Law of Gravitation, 184-5

Nitrogen atom split up by Rutherford, 43

Nomenclature, reforms needed in scientific and philosophical, 6

Nucleus, the: of the atom, 39, 40, 41, 42, 49; spontaneous breaking up in Radioactivity, 42; of cell, 62, 65; its part in heredity, 63; in cell-division, 70-71, 73

Object, relation of Subject and, 238-40

Objects: their fields, 17, 18-19, 112, 327-8; regarded as events in Space-Time, 111; misinterpreted by our intellect and senses, 111-12

Ontogeny repeats phylogeny, 74, 115

Organic and inorganic: vanished fixity of, 23; the colloidal surface as the bridge between, 47, 58; different structure of organic and inorganic compounds, 47-8, 64; the cell the real distinction between, 64-5

Organic Descent: its operation compared to that of Radioactivity, 52-3, 54. *See* Descent.

Organisms: power of regeneration possessed by, 80, 299; as typical wholes, 82-4, 96-7, 98, 101, 104, 110, 121-5, 127, 209, 214-15, 268, 271; relations of the parts and the whole in, 82-3, 84, 96-7, 101, 104, 106, 124-5, 127, 142, 210; inner co-ordination and self-regulation of, 97, 98, 106, 140-43, 159, 162, 177-8, 209, 210, 214-15, 216, 230-31, 234; their fields, 113-15; time factor in development and explanation of, 114-15, 210; transformation of a stimulus into free action by, 127, 135-6, 306-7; creativeness of wholes as seen in, 128-31, 132-5, 137; and their environment, 136-7, 138, 210, 238, 300, 340-41; Vitalism and, 159-61; selectiveness the fundamental property of, 162, 163; the freedom of, 162, 305, 306, 307; the laws of thermodynamics and the principles of life and mind in, 163-70; variations in, 208-9, 210-11 (*See* Variation); holistic repression of variations in, 215-16; individuality of, 232, 233, 242; development of mind in, 235-8; material objective of, 294; treated as synthetic units, 320; organic situations distinguished from, 339-40

Organs, holistic, 143

Origin of Species (Darwin), 23, 186-7

Osmosis, 65, 68, 76

Oxygen in the vital processes, 72

Pangenesis, Darwin's theory of, 188

Panpsychism and creative Evolution, 334-5

Past, the, in the activities of mind, 254-5, 257, 336-7

Pavlov, Professor, 205 *n.*

"Peraction," suggested term for body-and-mind relation, 270

Periodic Law, the, 38

Periodic Table, the, 41, 43, 54

Persona in Roman law, 282-3

Personality, the: the supreme embodiment of Holism, 105, 106, 140, 152, 153, 233, 246, 263-4, 276, 277, 283, 292, 293, 304, 308, 320; repressive activity of Holism and, 216; mind and the development of, 229, 233, 235, 264, 267, 274; Purpose a function of, 235, 240, 241, 259, 296; an apparent deviation from the main plan of Holism, 241; its present imperfection, 241, 246, 297, 299, 309-10; its basis universal, 245-6, 263-4; the hereditary past and, 255, 273, 274, 275, 276; as a whole, 263-89; the body and, 264-6, 279, 280; body and spirit in, 264-7; body and mind in, 267-71, 272, 273, 277; its own creative holistic activity, 271-3, 284, 292, 299, 304; its individuality, 274-5, 278-9, 284-5; constant and progressive, 275-6; as the Subject of experience, 277-8; psychology and, 277, 278, 279-82, 293; the subconscious mind and, 279-80, 296-7; the need for a science of, 281-2, 284-9, 293-4; evolution of the idea of, 282-3; value of biography to the study of, 284-9, 294; functions and ideals of, 292-316; an organ of self-realisation, 293, 294, 295, 315-16; ethical ideals of, 294, 295, 298, 299-300, 303, 305, 311, 312, 313, 314-16; the will and, 294, 295, 298, 309; the intelligence and, 294-5; inner control and direction of, 296-9, 300, 310; self-healing power of, 299-300; environment and, 300, 301-4; purity or wholeness of, 303-4, 305, 312-15; freedom of, 304, 309-12, 315, 316; and the idea of a Supreme Whole, 341-2

Personology, the science of Personality, 282, 284-9, 293-4

Phædo (Plato), 102

Philosophy, 90, 91, 297; its co-operation with science ensured by acceptance of creative Evolution, 90-92; the idea of the whole neglected by, 100; "the whole" in absolutism, 100, 102; the concept of the whole more precise than that of life in, 109-10; Holism and the old concepts of, 148; and the relations of body and mind, 156, 269; Personality in, 277, 278

Photo-synthesis in plants, 68, 69, 71, 76

Phylogeny repeated in ontogeny, 74, 115

Physical mixtures and chemical compounds, analogies from, 122-4, 128, 133, 134

Physical science: new concept of matter in, 3, 5, 51; recent progress in, 5, 7; the general acceptance of Evolution and, 11; and the relations of life and mind, 16, 154-5, 156, 157, 163-70; the error of abstraction in, 20; the doubtful validity of the concept of force in, 160; Holism and the Naturalism of, 329-31

Physico-chemical mechanisms, 150, 157, 173; Vitalistic hypothesis of,

159; structural equilibrium in, 173-4, 176; inner holistic character of, 174; the material of the new structure of life, 174-5, 176

Physics, the New, 37, 327; discoveries as to the constitution of matter, 37, 38-41, 327; assimilates chemical categories to physical, 47-8, 150; associates energy and mass, 164

Physiology: and Mechanism, 152; new categories demanded by, 177; and Personality, 279

Planck, Max, 39

Plants: the plant cell, 62, 65-9; the origin of the cell, 70-75; the process of reproduction similar to that of animals, 70, 71, 73; common origin of animals and, 75; causes of divergence of plant and animal forms, 75-7; holistic co-operation and co-ordination in, 78, 82, 83, 105, 106, 230, 231; regenerative power of, 80; ecological modifications and Variation in, 207, 208

Plato, 102

Pleurococcus, cell aggregation in, 75-6

Potentiality and organic creativeness, 131-2

Pre-established harmony, 220, 269-70, 272, 333, 334

Proton, the, 40, 41

Protoplasm, 46, 54, 62, 65, 114; enzyme action in, 46, 47, 68; its movement in the cell, 65; always in a process of creative change, 66-7; its formation of new protoplasm, 67; metabolism of, 67, 68; possible origin of its primitive forms, 72

Psychical nature of Natural and Sexual Selection, 13, 222-3

Psychological Principles (Ward), 280-81, 282

Psychology, 239, 244, 278; its methods, 227, 228, 278-80; the standpoint of Relativity essential to, 239, 240; natural science reunited with, 240; and the syntheses of Reason, 244; the services of Holism to, 259-60; and the Personality, 277, 278, 279-82, 293

Psycho-physical wholes, 154, 168, 169; the theory of Entelechy and 171-2

Purity, an ideal of the Personality, 303-4, 305, 312-15

Purpose: purposiveness distinctive of wholes, 139, 140; a function of Personality, 235, 240, 241, 259, 296; the highest manifestation of mind, 258-9; the purposive view of Evolution, 341-2, 343

Quantum, the, 39, 41; quanta of radiation released by electrons, 39-40, 49; discontinuity of the quanta, 40, 331; supremacy of the quantum law, 51

Racial and individual development, differentiation of, 198-9, 201-3, 204

Radioactivity, 23, 37, 38; a spontaneous breaking up of matter, 42; and the transmutation of elements, 42-3, 52, 53; its operation compared to that of Organic Descent, 52-3, 54

Radium converted into Lead, 42

Reality, 4; its nature obscured by the analytical character of thought, 15, 19, 111; the sensible order and the conceptual order, 49-50, 92, 111; the old static view of, 88, 89; creative Evolution and, 89, 92, 117, 132; not explained by Bergson's Duration, 92, 93, 94; Holism and the evolution of, 106-7, 108, 116, 142-4, 221; Relativity and, 111, 114, 122, 326; formula for its fundamental problems, 147, 149; individuation and, 239, 240, 242, 246; Reason and, 243; our sense of, and a sixth holistic sense, 256; Personality and, 263, 273, 277

Reason, 243-4; creative of values, 90, 243

Reflexes and the development of mind, 237, 249

Regeneration, organic, 80

Relativity, theory of, 16, 23, 24-5; its mathematical origin, 25-7; gravitation in, 25, 28-9, 30-31, 32, 185; the Space-Time universe, 28-32, 33, 326, 327; its subjec-

tive and objective aspects, 33-4; value of the Space-Time integration to our understanding of Nature, 111, 114, 122, 326; psychology and, 239
Reproduction, the process of: similar in plants and animals, 70, 71-3; cell-division, 70-71, 72-3; cell fusion, 73; reduction division, earlier than the separation of plant and animal forms, 73-4, 75; its holistic nature, 81-2; and the germ-cell theory of Variation, 198, 200
Rock-formations, measurement of age of, 42
Roman law, concept of personality in, 282-3
Röntgen, 39
Roux, Wilhelm, 189
Rutherford, Sir Ernest, 39, 43

Saltus, creative leaps in Evolution, 154, 197, 271
Sartor Resartus (Carlyle), 286
Scheiden, 62
Schwann, 62
Science, 247, 278; rigidity of its former concepts, 9, 16-17; linked with philosophy by creative Evolution, 90-92, 148; mechanistic outlook of, 91, 100, 103, 153, 322, 324; its neglect of the idea of the whole, 100, 322; Holism in relation to, 109-10, 142, 148, 160, 221, 248, 321-5; its mistaken view of life and mind as separate entities from the body, 154-5, 156, 157, 161, 324-5; the system of Science the greatest achievement of mind, 247-8, 249, 250; the body rehabilitated by, 266
Selectiveness: of matter, 47, 57, 161-2, 235; of life, 65, 79, 162, 163, 165, 167, 168, 170-71, 176, 235, 237, 249; of intellect, in Bergson's philosophy, 93-4, 111, 112; inherently holistic, 162-3, 343; the tap-root of will, 162, 163; of mind, 167, 168, 171
Self, the, 292; as the centre of experience, 238, 240, 243, 245-6, 292-3; an apparent rebel against the universal order, 241, 244, 246; as the centre of the higher order of Holism, 243, 293; largely a social construction, 244-5, 246. *See* Personality.
Self-determination, the true ideal of human development, 309, 311
Self-direction in life and mind, 162, 163-4, 165, 167, 168, 170-71, 176-7, 250-52, 270
Senses, the: their limitations and defects, 111-12; holistic character of, 255-6, 284; mind and, 256-7
Sexual Selection: Darwinian theory of, 12-13; its operation limited to males, 13; emotional sensitiveness of females implied by, 13, 222; psychical nature of, 13-14, 222-3
Shakespeare, his hidden Personality, 288
Single-content cell, generation of, 74
Society: holistic, 106, 344; creation of, by mind, 251, 252; Personality and, 296; group fields and, 339-40, 342
Socrates, 229
Solids, internal structure of, 45
Soul, the: 19th-century materialism and, 8, 9; as a whole, 102; mistaken physical analogies of, 158; relations of body and, 161, 264, 265-7; Personality and, 264, 265-7, 298; and the ideals of Holism, 312, 314, 315, 316, 335-6, 337-8; the universe not a Society of Souls, 335-6, 337; environment and, 336-7
Space: in the theory of Relativity, 26, 27, 28-32, 33, 34; as conceived by Newton and Kant, 32-4, 94
Space-Time universe, the, 27-32, 38, 41, 173, 326, 327; value of the concept to our understanding of Nature, 111, 114, 122, 173
Special Theory of Relativity (Einstein), 26-7
Spectrum analysis, 39, 40
Spencer, Herbert, 188, 202
Spinoza, 270, 272, 321 *n.*, 334-5
Spirit, the: spiritual ideals of Holism, 99, 106, 107, 108, 139, 144, 152, 222, 240, 241, 242, 243; spiritual structure of man, 152; materialist misconceptions of the spiritual, 158; the trend of Evolution towards, 178, 181;

relations of body and, 264, 266-7; Personality best studied in spiritual life, 286-8; spiritual objective of Personality, 294, 299, 305

Spiritual idealism (Spiritualism): its battle with materialism over Evolution, 8-9; life and mind in, 169, 170; irreconcilable with creative Evolution, 330-31, 332, 333-6, 337, 343

Sporophyte generation, the, 74

Sports, biological, and Personality, 276

State, the, a super-individual whole, 98, 106

Structure: structural character of the universe, 24, 31, 34, 37, 44, 230, 323; of matter, 37-42, 43-4, 51, 55, 56, 95, 162, 173-4, 230, 231, 331; relations of energy and, 41, 44, 51, 55, 56, 327; internal structure and external properties, 43-5, 112, 173-4, 176; dynamic self-controlled equilibrium of, 43-5, 173-4, 176-9, 235; organic, 64, 114, 169, 177-8; modern science and the study of, 90-91; and the process of creative Evolution, 91, 92, 129-31, 133-5, 143-4, 169-70, 178, 230, 323; mind and, 94, 95, 252, 253; intellect and experience and, 94-5; functions and structure in wholes, 104, 105, 106, 107, 112, 122-4, 153; holistic character of, 104, 105, 106, 107, 108, 122, 123-4, 127, 153, 174, 175, 178, 179, 180, 232; the field and, 112-15; life a new structure based on the physico-chemical structures of Nature, 174-6, 178, 179, 181, 230, 335; mind as a new departure in, 178, 181, 229, 231, 235-6, 253; distinction between Mechanism and, 180-81; hereditary, less important in the human than the animal, 252, 253; unit character of, 331

Struggle for existence, the, 11-12, 13, 14-15, 186, 187-8, 189, 206; the exception, not the rule, 218, 219

Subconscious mind, the, 253-5; in the Personality, 279-80, 296

Subject-Object relation, the, 238-40

Substance, divine, Spinoza's, 270, 272

Sunlight and its action upon chlorophyll, 47, 58, 68, 71, 76-7

Survival of the fittest, law of, 11-12 13, 15, 186, 187-8, 189

Synthetic unity of wholes, the, 123-5, 126-7, 129, 130

Teleological view of the universe, 341-2, 343. *See* Purpose.

Telepathy, 257

Thermodynamics, laws of, 155; and the principles of life and mind, 163-9

Things and their fields, 17-19, 112, 113-14; 327-8; as events in Space-Time, 111; misinterpreted by the senses, 111-12

Thomson, Sir J. J., 39

Thorium, its conversion into Radium, 42

Thought: baffled by misconceptions of life, mind, and matter, 2, 3, 4, 157, 158-9; value of the idea of fields, 17-19, 31; errors due to analytical character of, 19-20; creativeness of, 90; deductive and inductive, 128-9; structural character of, 180; a social instrument, 245; telepathy and, 257

Time: in the theory of Relativity, 26, 27, 28-32, 33, 34; as conceived by Newton and Kant, 32-4, 94; and Bergson's principle of Duration, 93; the Time factor in the field of mind, 254. *See* Space-Time universe.

Transmutation of elements, 42-3, 52, 53

Treviranus, 62

Tropisms, 236, 237, 249

Unicellular organisms, reproduction of, 70-71, 73, 75-6

Unity: of structure and functions in the whole, 84, 122-5, 127, 141-2; Holism the tendency towards, 179, 181, 220, 232

Universality and individuation, two tendencies of Holism, 238, 240, 241, 242; Reason the organ of universality, 243-4

Universe, the: its structural character, 24, 31, 34, 37, 44, 230, 323, 327; stationariness in, an illusion, 25-6, 31; the Space-Time uni-

INDEX

verse, 27-32, 38, 41, 111, 114, 122, 173, 326, 327; Action its inmost nature, 31, 41, 51, 56, 326-7, 328; its creativeness, 55-7, 89, 132; Holism fundamental in, 82, 84, 97-8, 99, 100, 107, 108, 143, 179, 181, 307-8, 319-45; two contrasted explanations of, 88-90; mechanistic view of, 89, 100, 103; Bergson on Duration as its creative principle, 92-3; absolutist view of, 100, 102; its friendliness, 218, 220, 343; the value of mind to, 248-9; transformed by Personality, 277-8, 320; Freedom the rule of, 307-8; Naturalistic view of, 329-30; Idealist view of, 330-31, 337; monadic view of, 333-4, 339; teleological view of, 341-2, 343; the assumption of a Supreme Mind in, 341-2

Uranium, 41: its conversion into Radium, 42

Use and routine, and modifications and variations, 203, 204, 206-8, 212, 219-20

Values, or creative Ideals of Holism, 107, 144, 221, 241, 243, 259, 305, 335, 344-5

Variation, 141, 192; Darwinian theory of, 186, 187, 188, 189, 192-4, 206; Neo-Darwinian theories of, 188, 190-205; the germ-cell theory of, 188, 190, 191, 194, 197-205; mechanistic view of, arbitrary and misleading, 190, 194-5, 209, 214, 215, 220; the principle of Holism and, 191, 206, 209, 210-15, 332, 342-3; Mutation theory of, 194, 196-7; Mendelism, 194, 195-6; possible influence of modifications on the germ-cell, 203-5, 206, 207, 212, 219; the natural selection of small variations, 205-9; environment and, 207, 208, 210, 218, 219; Holistic Selection and, 211, 212-15; Holism and the repression of variations, 215-16

Vitalism, the hypothesis of, criticised, 159-61; the theory of Entelechy in, 171-2

Wallace, A. R., 12
War, the Great, 344

Ward, Professor James, 280-81, 282, 334, 335

Weismann, 14; his germ-cell theory of Variation, 190, 191, 194, 197-205; his doctrine of germinal isolation, 198-9, 201, 202, 203, 219, 220

Whole, the: the reciprocal influence of the whole and its parts, 77-84, 107, 122-7, 138-9, 209-10, 214, 271, 272-3, 341; fundamental tendency of the universe towards wholes, 82, 84, 97-8, 99, 100, 107, 108, 143, 179, 181, 307-8, 319-21, 326, 329, 335, 337-8, 344-5; the character of wholes, 98, 99, 101-2, 103-4, 106-7, 108, 121, 122-3, 341; ideal wholes, 98, 105-6, 107, 144, 222, 243, 259, 305, 311, 312, 313, 314-16, 329, 344-5; the whole in absolutist philosophy, 101-2, 103; creativeness of wholes, 101, 102, 104, 128-32, 133-6, 142-4, 221, 222, 271, 273, 304; relation of function and structure in wholes, 104, 106, 107, 112, 122-4; progressive scale of wholes, 105, 106-7; the concept of life and that of the whole, 109-10, 160-61; wholes and their fields, 110-16, 335, 339, 340; stimulus transformed into free action by, 127, 135-6, 138, 306-7; the whole and the idea of cause, 126-7, 135-6, 137, 138; freedom of wholes, 137-8, 306-8, 309-12; individuality of wholes, 139-40, 232-3; co-ordination and co-adaptation in wholes, 140-42, 208-9, 214-15, 220, 333-4; selectivity of wholes, 162-3; the trend towards a greater Whole, 179, 243, 293, 315, 316; Personality as the highest whole, 263, 264, 267, 272, 273, 275, 292, 294, 295, 297, 302, 304, 309, 310; psychic wholes, 271, 273; monads and wholes, 332-4, 335, 339; the nature of the Supreme Whole, 338, 341; Nature as a society of wholes, 340, 342, 343

Wholeness: in life, 77-84, 97; the aim of Personality, 295-6, 297, 298, 300, 302-4, 309, 312-13, 314-15, 316

Will, the, 155, 250; selectivity and,

162, 163; development of, 240, 243; the basis of Personality, 294, 295, 296, 298; freedom and, 307, 309, 310

Wordsworth, 336

X-rays and the investigation of atomic structure, 39, 112